Analysis and Linear Algebra

Thomas Holey · Armin Wiedemann

Analysis and Linear Algebra

An Introduction for Economists

 Springer

Thomas Holey
Duale Hochschule Baden-Württemberg Mannheim
Mannheim, Germany

Armin Wiedemann
Duale Hochschule Baden-Württemberg Mannheim
Mannheim, Germany

ISBN 978-3-662-66246-5   ISBN 978-3-662-66247-2 (eBook)
https://doi.org/10.1007/978-3-662-66247-2

© The Editor(s) (if applicable) and The Author(s), under exclusive license to Springer-Verlag GmbH, DE, part of Springer Nature 2023

This work is subject to copyright. All rights are solely and exclusively licensed by the Publisher, whether the whole or part of the material is concerned, specifically the rights of translation, reprinting, reuse of illustrations, recitation, broadcasting, reproduction on microfilms or in any other physical way, and transmission or information storage and retrieval, electronic adaptation, computer software, or by similar or dissimilar methodology now known or hereafter developed.

The use of general descriptive names, registered names, trademarks, service marks, etc. in this publication does not imply, even in the absence of a specific statement, that such names are exempt from the relevant protective laws and regulations and therefore free for general use.

The publisher, the authors, and the editors are safe to assume that the advice and information in this book are believed to be true and accurate at the date of publication. Neither the publisher nor the authors or the editors give a warranty, expressed or implied, with respect to the material contained herein or for any errors or omissions that may have been made. The publisher remains neutral with regard to jurisdictional claims in published maps and institutional affiliations.

Responsible Editor: Claudia Rosenbaum
This Springer imprint is published by the registered company Springer-Verlag GmbH, DE, part of Springer Nature.
The registered company address is: Heidelberger Platz 3, 14197 Berlin, Germany

Preface to the Fifth Edition

In the fifth edition, we have considered the treatment of sequences and series at some points. With this we take into account the content-related structure of many lectures and want to express this with the new title *Analysis and Linear Algebra*. The subtitle *An Introduction for Economists* refers to the applications and examples that are to be assigned to economics or economic informatics.

The chapter Applications of Differential Calculus has been extended by a section on Taylor series, in examples it is shown how these tools can be used. In the chapter Financial Mathematics, we have taken into account the trend of the times and the interest rates have been approximated to the current values. Some more exercise problems have been added and literature references have been added.

The slide set as a lecture basis and the detailed solutions to the exercises for the 5th edition are available on Springerlink.

Mannheim, Germany Thomas Holey
May 2021 Armin Wiedemann

Preface to the Fourth Edition

In the fourth edition, errors that affected the figures in Chap. 6 were corrected.

The interest in exercise problems is still great. Accordingly, we have taken on additional exercise problems. The note on the URL of the publisher's website, where detailed solutions to the tasks are available, is now directly in the exercise part of the individual chapters. The literature references have been updated again.

Mannheim, Germany Thomas Holey
July 2015 Armin Wiedemann

Preface to the Third Edition

With the third edition we have complied with the request of some readers to include a chapter with solutions to the exercise problems. In this chapter the solutions to all exercise problems are given, partly the solution path is briefly sketched. Detailed solution paths to the exercise problems can still be accessed via the website of the publisher for the book. The literature references have been revised.

Mannheim, Germany
October 2012

Thomas Holey
Armin Wiedemann

Preface to the Second Edition

In the second, corrected and revised edition, we have taken up numerous suggestions from lecturers and students who use the book for lectures and exercises. We have responded to the request to provide more exercise problems. So after each chapter there are some basic exercise problems as well as further problems from the field of business applications. The solutions to the problems can be accessed again via the corresponding link of the Springer-Verlag.

We would like to thank all those who have given us hints and suggestions for the second edition. In particular, we would like to thank Ms. Dipl. Math. Eva Schmitt-Leiß and Mr. Prof. Dr. Klaus Gläser for helpful discussions.

Mannheim, Germany
August 2009

Thomas Holey
Armin Wiedemann

Preface to the First Edition

Quantitative methods form an important basis in almost all economic disciplines. Accordingly, mathematical introductory lectures can be found in the framework study plans of these degree programs.

The present book *Mathematics for Economists* in the series BA-Kompakt is very strongly oriented towards the framework study plan for the field of economics at vocational academies in Baden-Württemberg. In the selection of material, we have endeavored to include the most important topics for students of business administration and business informatics. A wide range of mathematical basics is presented, which are often also the subject of upper secondary level: functions of a variable, differential and integral calculus. The book offers the lecturer the possibility to repeat these basics very quickly and to concentrate more on the applications, depending on the students' level of knowledge. If it appears more reasonable and necessary to spend more time on the basics, one can restrict oneself to the special case of two variables when dealing with functions of several variables. Then, in linear algebra, the concepts of determinant and eigenvalue of a matrix are not needed.

In this way, a certain flexibility is created without completely omitting topic areas of the framework study plan. Some basics are listed in the first chapter, which should be part of the "toolbox". The book offers students the opportunity to test their knowledge with a short self-test.

We would like to thank Mrs. Prof. Dr. Irene Rößler and Mr. Prof. Dr. Frank Hubert for many helpful discussions and suggestions. We would like to thank the editors, Prof. Dr. Martin Kornmeier and Prof. Dr. Willy Schneider, for including the book in the series BA-Kompakt. Our thanks go to Mrs. Katharina Wetzel-Vandai and Mrs. Gabriele Keidel at Springer-Verlag for their helpful support in editorial matters.

Finally, we would like to point out that a set of slides is available as a lecture basis and the solutions to the exercises can be downloaded from the publisher at the URL
  http://www.springer.com/978-3-7908-1973-1. The solutions to the test are given in the appendix.

Mannheim, Germany Thomas Holey
Juni 2007 Armin Wiedemann

Contents

1 Elementary Basics 1
  1.1 Elementary From Propositional Logic 1
  1.2 Set Theory 4
  1.3 Basic Arithmetic Operations 9
  1.4 Equations 16
  1.5 Trigonometry 21
  1.6 Test 22
  References 24

2 Functions 25
  2.1 Definition and Representation of Functions 25
  2.2 Some Elementary Functions 28
    2.2.1 Linear Function 28
    2.2.2 Quadratic Function 30
    2.2.3 Integer Rational Functions or Polynomials 31
    2.2.4 Power Function 31
    2.2.5 Fractional Rational Functions 32
    2.2.6 Hyperbolic Function 33
    2.2.7 Root Function 33
    2.2.8 Exponential Function 34
    2.2.9 Logarithm Function 36
    2.2.10 Trigonometric Functions 37
    2.2.11 Sectionally Defined Functions 41
    2.2.12 Some Economic Functions 42
  2.3 The Inverse Function 47
  2.4 Chained Functions 51
  2.5 Properties of Functions 52
    2.5.1 Limitation 52
    2.5.2 Monotonicity 53

		2.5.3	Symmetry	54
		2.5.4	Injectivity, Surjectivity and Bijectivity	55
	2.6	Limits		57
		2.6.1	Convergence and Limits of Sequences and Series	57
		2.6.2	The Limit Concept for Functions	67
		2.6.3	Cauchy's Definition of the Limit of Functions	70
		2.6.4	Limit Considerations of Some Elementary Functions	71
		2.6.5	Calculation Rules for Limits	73
		2.6.6	Examples of Limit Considerations	74
	2.7	Continuity of Functions		79
	2.8	Exercises		82
	References			87
3	**Differential Calculus**			89
	3.1	The Concept of the Derivative		89
	3.2	Derivatives of Elementary Functions		93
	3.3	Derivative Rules		96
	3.4	Differentiability		100
	3.5	Higher Derivatives, Extreme Values and Turning Points		103
	3.6	Applications of Differential Calculus		107
		3.6.1	L'Hospital's Rule	107
		3.6.2	Determination of Zeros with the Newton Method	109
		3.6.3	Taylor Series	115
		3.6.4	Curve Discussion	122
		3.6.5	Limit Functions	127
		3.6.6	Elasticity of Functions	128
	3.7	Exercises		130
	References			133
4	**Integral Calculus**			135
	4.1	The Indefinite Integral		135
		4.1.1	Primitives of Elementary Functions	136
		4.1.2	Linearity of the indefinite integral	137
	4.2	The Definite Integral		138
		4.2.1	Properties of the Definite Integral	141
		4.2.2	Value of an Integral	143
		4.2.3	Area Between Two Curves	145
		4.2.4	Improper Integrals	147
		4.2.5	Partial Integration	148
		4.2.6	Integration by Substitution	150
	4.3	Application of Integral Calculus		151
		4.3.1	Determination of the Economic Function from the Marginal Function	151

	4.3.2	Consumer Rent.		152
	4.3.3	Producer Surplus		153
	4.3.4	Numerical Integration		154
4.4	Exercises			157

5 Linear Algebra 161
- 5.1 Vectors 161
  - 5.1.1 Definition of Vectors 161
  - 5.1.2 The Linear Combination of Vectors..................... 164
  - 5.1.3 Scalar Product of Two Vectors......................... 166
- 5.2 Matrices 167
  - 5.2.1 Definition of a Matrix 167
  - 5.2.2 Addition of Matrices 170
  - 5.2.3 Multiplication by a Scalar 171
  - 5.2.4 Matrix Multiplication.................................. 171
  - 5.2.5 Calculation Rules of the Matrix Product 175
  - 5.2.6 Inverse Matrix 177
- 5.3 Systems of Linear Equations 178
  - 5.3.1 Basic Considerations 178
  - 5.3.2 Solution Methods for Systems of Linear Equations 182
  - 5.3.3 Standardized form of Systems of Linear Equations 189
  - 5.3.4 Matrix Inversion....................................... 190
  - 5.3.5 Business Applications 194
  - 5.3.6 Eigenvalues of a Matrix................................ 196
- 5.4 Exercises 198
- References... 202

6 Functions with Several Variables................................ 205
- 6.1 Introduction and Representation 205
- 6.2 Differential Calculus for Functions with Several Variables. 210
  - 6.2.1 Partial Derivative 210
  - 6.2.2 The Total Differential................................. 212
- 6.3 Extremum Values of Functions with Several Variables 215
  - 6.3.1 Extremum without Boundary Conditions 215
  - 6.3.2 Extremum Values with Boundary Conditions 222
- 6.4 Exercises 228
- References... 231

7 Financial Mathematics 233
- 7.1 Interest Calculation 233
  - 7.1.1 Simple Interest.. 233
  - 7.1.2 Compound Interest...................................... 235

		7.1.3	Annuity Calculation	237
		7.1.4	Yearly Interest	238
	7.2	Repayment Calculation		240
	7.3	Exercises		247

Appendix A 249

References 279

Elementary Basics

Learning Objectives (This Chapter Provides)

- Basics from various areas of mathematics
- A summary of the topics that are required for understanding the book
- A self-assessment through a test ◄

1.1 Elementary From Propositional Logic

The **proposition** is a fundamental concept of mathematics.[1] Essential aspects of propositional logic were developed by Aristotle (384–322 BC).

▶ **Definition (Description of a Proposition)** A proposition is a linguistic construct for which it makes sense to ask whether it is true (t) or false (f).

Propositions are, for example:

3 is an odd number.
5 is less than 3.

---

[1] Introductions to propositional logic can be found in the monographs Kelly (2003), Kreuzer and Kühling (2006), Staab (2012) or Winter (2001).

© The Author(s), under exclusive license to Springer-Verlag GmbH, DE, part of Springer Nature 2022
T. Holey and A. Wiedemann, *Analysis and Linear Algebra*,
https://doi.org/10.1007/978-3-662-66247-2_1

Questions and requests are therefore not propositions. A **proposition form**[2] contains a variable; depending on the value of this variable, the proposition form turns into a true or false statement. The values of the variable for which a proposition form turns into a true statement are called the solutions of a proposition form.

proposition forms can be **unsolvable** if there is no solution and **universal** if every value from the possible value set satisfies the proposition form.[3]

Connections of Propositions (Conjunctions)
The proposition logic deals with the connection of statements.

The **negation** is a unary connection, it reverses the truth value of a statement. For the negation of a proposition A we write $\neg A$. The truth value table of the negation looks as follows:

A	$\neg A$
t	f
f	t

Of the two-digit connections we consider the **conjunction** (also AND-connection) $\wedge$ and the **disjunction** (also OR-connection) $\vee$. The truth value table of these connections results in:

A	B	$A \wedge B$	$A \vee B$
t	t	t	t
t	f	f	t
f	t	f	t
f	f	f	f

The conjunction of two propositions is only then true if both propositions are true, while the disjunction is true if at least one of the propositions is true. One also speaks here of the *inclusive* OR in contrast to the *exclusive* OR, in which exactly one of the two propositions is true.

Example

A: 3 is an odd number.
B: $3 \geq 6$.

---

[2] In the context of mathematics.
[3] Universal statement forms are also called **tautologies**.

1.1 Elementary From Propositional Logic

Then:

$A \wedge B = $ f (since B is false)
$A \vee B = $ t (since A is true).

▶ **Definition (Logical Inference (Implication))** A proposition B is called a logical inference of a proposition A if the following holds: If A is true, then B is also true: $A \Rightarrow B$ (Read: If A, then B or A implies B.).

The implication is often expressed as follows:

$$A \text{ is a sufficient condition for } B$$

or:

$$B \text{ is a necessary condition for } A.$$

The implication allows for B to be true even if A is false. For proposition forms $A(x)$ and $B(x)$, this means: $A(x) \Rightarrow B(x)$, if all solutions of $A(x)$ are also solutions of $B(x)$. $B(x)$ can however have solutions that are not solutions for the proposition form $A(x)$.

Example Consider the two propositions:

$A(x):$ $\quad x = 3$ $\qquad$ Solution : $x = 3$
$B(x):$ $\quad x^2 = 9$ $\qquad$ Solutions: $x = 3$ and $x = -3$
$\qquad\qquad\qquad\qquad\qquad x = 3 \implies x^2 = 9.$

▶ **Definition (Equivalence)** The two propositions A and B are called equivalent if the following is true: If A is true, then B is true, and if B is true, then A is true. $A \Leftrightarrow B$ (Read: B if and only if A or A is equivalent to B).

Two propositions forms $A(x)$ and $B(x)$ are called equivalent if $A(x) \Leftrightarrow B(x)$, if all solutions of $A(x)$ are also solutions of $B(x)$ and vice versa, all solutions of $B(x)$ are also solutions of $A(x)$.

Example

$$2x = 5 \quad \Longleftrightarrow \quad x = \frac{5}{2}.$$

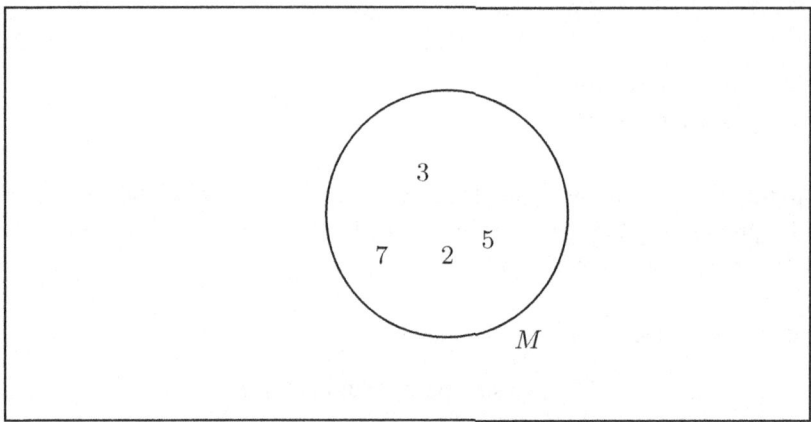

Fig. 1.1 A simple set representation as a Venn diagram

1.2 Set Theory

The concept of set is elementary for the description of mathematical relationships.[4] The basis of the concept of set is the following definition, which goes back to Georg Cantor (1845–1918):

▶ **Definition (Set)** A **set** is a collection of certain, well-defined objects of perception or thought—which are called the **elements** of the set—to a whole.

For the membership of an object to a set M the symbol $\in$ is usually used: $x \in M$, x is an element of the set M. Analogously, $x \notin M$ means that x is not an element of the set M. If a set contains no elements, this is called the **empty set** and is written as $\emptyset = \{\ \}$.

There are various ways to represent sets:

- The enumerative form: $M = \{2, 3, 5, 7\}$.
- The descriptive form: $M = \{x \,|\, x \text{ is a prime number } \leq 7\}$.
- The representation in the form of a Venn diagram[5] as in Fig. 1.1.

---

[4] Introductions to set theory can be found in Dean (2003), Deiser (2010), Garnier and Taylor (2002), Gregg (1998), Koshy (2004), Toenniessen (2019) or Winter (2001).

[5] These diagrams were introduced by the English logician and philosopher John Venn (1834–1923) to describe sets in a visual way.

1.2 Set Theory

Sets of numbers play an important role in this book.[6] In particular, we consider:

- The set of natural numbers

$$\mathbb{N} = \{1, 2, 3, \ldots\}.$$

Occasionally we consider the set of natural numbers together with the number 0, we denote this set as

$$\mathbb{N}_0 = \{0, 1, 2, 3, \ldots\}.$$

- The set of integer

$$\mathbb{Z} = \{\ldots, -3, -2, -1, 0, 1, 2, 3, \ldots\}.$$

- The set of rational numbers

$$\mathbb{Q} = \left\{ \frac{p}{q} \,\middle|\, p, q \in \mathbb{Z}, q \neq 0 \right\}.$$

- The set of real numbers we denote by $\mathbb{R}$. The set of rational numbers $\mathbb{Q}$ is extended because certain quantities such as the value of π or $\sqrt{2}$ cannot be expressed by quotients of two integers. The set of positive real numbers is denoted by $\mathbb{R}^+$.

Subsets

Under a **subset** $T \subseteq M$ (read: T is a subset of M) one understands the set that fulfills the condition: $x \in T \Rightarrow x \in M$, thus every element of T is also an element of the set M.

Example The set $T = \{2, 3\}$ is a subset of $M = \{2, 3, 5, 7\}$.

Subsets of real numbers are called open or closed intervals: For $M = \mathbb{R}$ the set $T_g = [a, b]$ is a closed interval and is defined as follows:

$$T_g = \{x \mid x \in \mathbb{R} \wedge a \leq x \leq b\}.$$

The open interval $T_o = \,]a, b[$ is given by:

$$T_o = \{x \mid x \in \mathbb{R} \wedge a < x < b\}.$$

Cartesian Product of Sets

The **Cartesian product** of the sets $M_1, M_2, \ldots, M_n$ is the set of all ordered n-tuples that can be formed from the sets:

$$M_1 \times M_2 \times \ldots \times M_n = \{(x_1, x_2, \ldots, x_n) \mid x_i \in M_i, i = 1, 2, \ldots, n\}.$$

---

[6] We cannot go into the strictly axiomatic construction of the number system here. See the readable presentation in Hilgert and Hilgert (2021), Appendix A, the book by Kramer and von Pippich (2013), or the recommended presentation by Körner (2020).

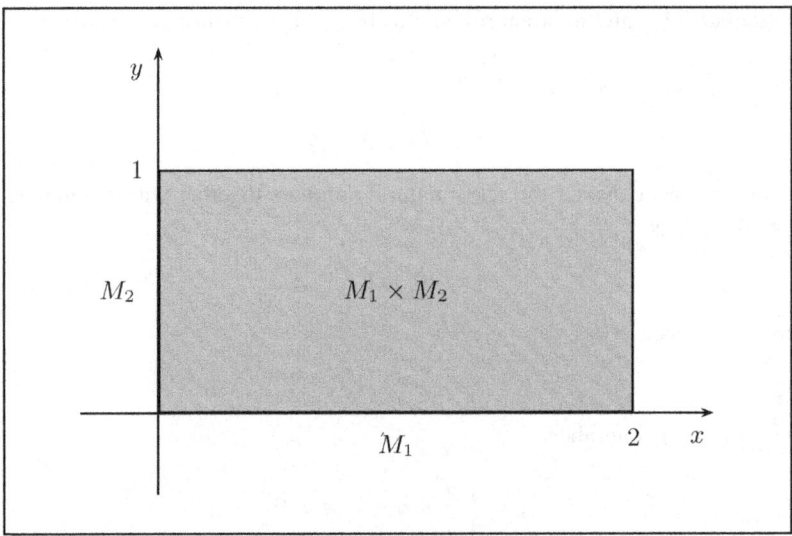

Fig. 1.2 Representation of the cartesian product of two intervals

Example Let $M_1 = \{2, 3, 4\}$ and $M_2 = \{a, b\}$, then

$$M_1 \times M_2 = \{(2,a), (2,b), (3,a), (3,b), (4,a), (4,b)\}.$$

Of particular interest are the Cartesian products of number sets. We consider the two intervals:

$$M_1 = \{x \mid x \in \mathbb{R} \text{ and } 0 \leq x \leq 2\}$$

and

$$M_2 = \{y \mid y \in \mathbb{R} \text{ and } 0 \leq y \leq 1\}.$$

Then the Cartesian product is:

$$M_1 \times M_2 = \{(x,y) \mid x, y \in \mathbb{R} \wedge 0 \leq x \leq 2 \wedge 0 \leq y \leq 1\}.$$

This cartesian product can be represented graphically as a rectangle in the plane (see Fig. 1.2).

The cartesian product

$$\underbrace{\mathbb{R} \times \mathbb{R} \times \ldots \times \mathbb{R}}_{n \text{ times}} = \mathbb{R}^n \quad (read : Rn)$$

describes the n-dimensional space, in particular, provides $\mathbb{R}^3$ the points in three-dimensional space (see Fig. 1.3).

1.2 Set Theory

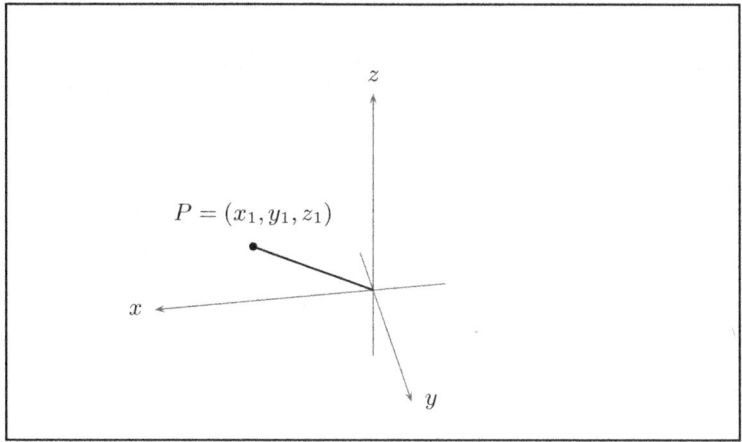

Fig. 1.3 The points of three-dimensional space are ordered triples $P = (x_1, y_1, z_1)$

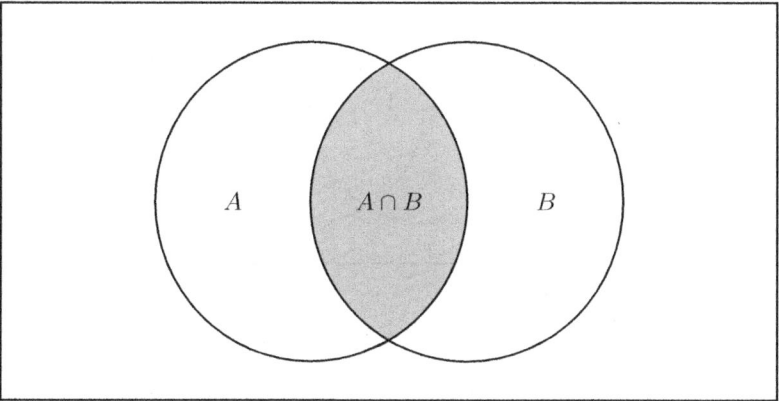

Fig. 1.4 The intersection of two sets $A \cap B$ as a Venn diagram

Set Operations

The intersection

$$A \cap B \quad (A \text{ intersection } B)$$

consists of those elements that are contained in A and in B (Fig. 1.4):

$$A \cap B = \{x \mid x \in A \ \wedge \ x \in B\}.$$

If the intersection $A \cap B$ is empty, then the two sets A and B are called **disjoint**.

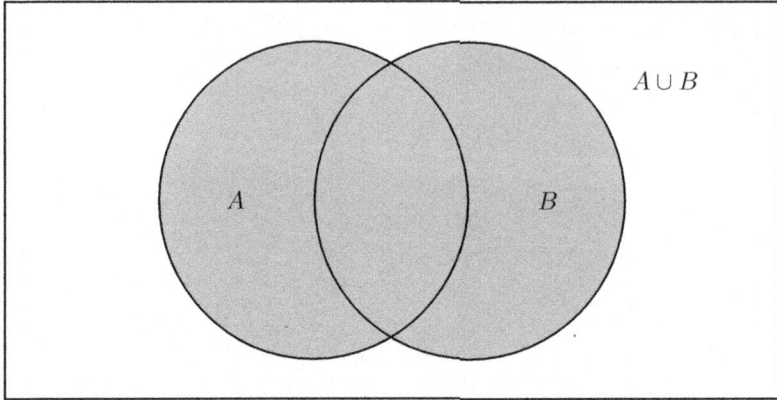

Fig. 1.5 Venn diagram of the union of two sets A and B

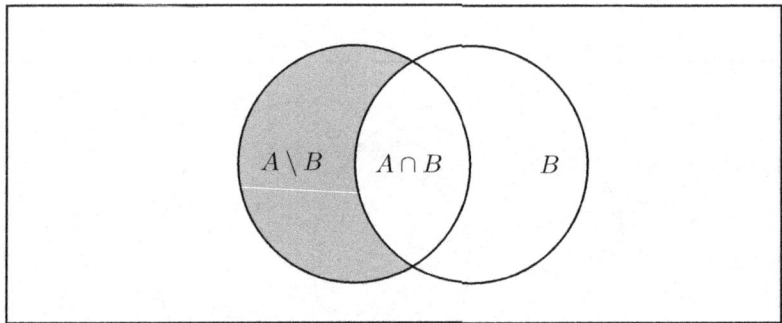

Fig. 1.6 The set difference

The **union** $A \cup B$ of two sets is the set of all elements that belong to A **or** to B, i. e. (Fig. 1.5):

$$A \cup B = \{x \mid x \in A \ \vee \ x \in B\}.$$

The **difference** $A \setminus B$ of two sets is the set of all elements of A that do not belong to B, i. e. (Fig. 1.6):

$$A \setminus B = \{x \mid x \in A \text{ and } x \notin B\}.$$

If $A \subseteq B$, then a **complementary set** of A can be defined as follows (Fig. 1.7):

$$\overline{A} = \{x \mid x \in B \wedge x \notin A\}.$$

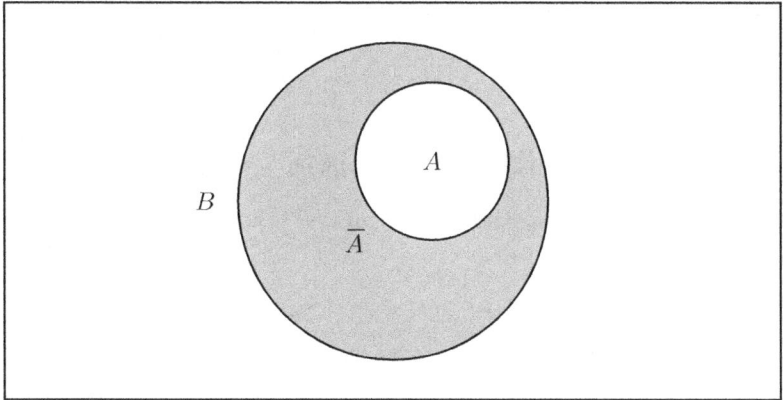

Fig. 1.7 Venn diagram for the complementary set

1.3 Basic Arithmetic Operations

The set of rational numbers and the set of real numbers form, with the operations of addition '+' and multiplication '·', an **algebraic structure** called a **field**.[7] The following axioms apply to fields, as can be easily verified by example.[8] We formulate the axioms for the real numbers:

1. The sum of two numbers $a, b \in \mathbb{R}$ exists for all a, b and is unique:

$$s = a + b \qquad (s \in \mathbb{R}).$$

2. The product of two numbers $a, b \in \mathbb{R}$ exists for all a, b and is unique:

$$p = a \cdot b \qquad (p \in \mathbb{R}).$$

It is said that $\mathbb{R}$ is **closed** under the two operations of addition and multiplication.

3. The **commutative laws** apply:

$$a + b = b + a \qquad \text{for all } a, b \in \mathbb{R}$$

and

$$a \cdot b = b \cdot a \qquad \text{for all } a, b \in \mathbb{R}.$$

---

[7] A very readable introduction to the topic of algebraic structures can be found in Basieux (2000).
[8] Axioms are unprovable basic assumptions that must be free of contradiction.

4. The **associative laws** apply:

$$a + (b + c) = (a + b) + c \qquad \text{for all } a, b, c \in \mathbb{R}$$

and

$$a \cdot (b \cdot c) = (a \cdot b) \cdot c \qquad \text{for all } a, b, c \in \mathbb{R}.$$

5. For each operation there is a **neutral element** in $\mathbb{R}$. The neutral element of addition is the $0 \in \mathbb{R}$ with

$$a + 0 = 0 + a = a \qquad \text{for all } a \in \mathbb{R}.$$

The neutral element of multiplication is the $1 \in \mathbb{R}$ with

$$a \cdot 1 = 1 \cdot a = a \qquad \text{for all } a \in \mathbb{R}.$$

6. There is an **inverse element** in $\mathbb{R}$ for each operation, of the type that the respective neutral element results:

$$a + (-a) = 0 \qquad \text{for all } a \in \mathbb{R}.$$

$(-a) \in \mathbb{R}$ is called additive inverse. The multiplicative inverse has the property:

$$a \cdot \frac{1}{a} = 1 \qquad \text{for } a \in \mathbb{R} \setminus \{0\}.$$

7. There is a **distributive law** of the form:

$$a \cdot (b + c) = a \cdot b + a \cdot c \qquad \text{for } a, b, c \in \mathbb{R}.$$

The inverse elements are used to reduce subtraction and division to addition and multiplication:

$$a - b = a + (-b)$$

$$\text{and} \qquad a : b = \frac{a}{b} = a \cdot \frac{1}{b}.$$

From these axioms, elementary calculation rules can be derived.

Fractional Arithmetic

Multiplication of fractions:

$$\frac{a}{b} \cdot \frac{c}{d} = \frac{a \cdot c}{b \cdot d} \qquad (b, d \neq 0).$$

Addition of fractions:

$$\frac{a}{b} + \frac{c}{d} = \frac{ad + cb}{b \cdot d} \qquad (b, d \neq 0).$$

1.3 Basic Arithmetic Operations

Power Arithmetic

A **power** of a real number $a \in \mathbb{R}$ is defined as

$$a^n = \underbrace{a \cdot a \cdot \ldots \cdot a}_{n \text{ Factors}} \qquad a \in \mathbb{R}, n \in \mathbb{N}.$$

The following calculation rules apply to powers:

$$a^m \cdot a^n = a^{m+n} \tag{1.1}$$

$$(a^m)^n = (a^n)^m = a^{m \cdot n} \tag{1.2}$$

$$(a \cdot b)^n = a^n \cdot b^n. \tag{1.3}$$

With the definitions

$$a^{-n} = \frac{1}{a^n} \qquad (a \neq 0; n \in \mathbb{Z})$$

and

$$a^0 = 1 \qquad \text{for all } a \in \mathbb{R}, a \neq 0$$

it follows from Eq. (1.1):

$$\frac{a^m}{a^n} = a^{m-n}. \tag{1.4}$$

The validity of the power calculation rules is thus extended to integers.

By defining the solution of the equation $x^n = a$ as

$$x = a^{\frac{1}{n}} = \sqrt[n]{a},$$

the concept of power can also be extended to rational exponents. The power calculation rules remain valid in general for $a>0$. In special cases, a definition for $a<0$ is meaningful, for example in: $\sqrt[3]{-8} = -2$.

Logarithm

The definition of the logarithm results from solving the equation

$$b^x = y$$

for x.

Definition

$$b^x = y \iff x = \log_b y$$

with
$$b \in \mathbb{R}^+ \setminus \{1\}, y \in \mathbb{R}^+, x \in \mathbb{R}.$$

The number b is called the **base** of the logarithm.
Computational rules for logarithms (let $x, y > 0$):

$$\log_b(x \cdot y) = \log_b x + \log_b y \tag{1.5}$$

and

$$\log_b(x^k) = k \cdot \log_b x, \quad k \in \mathbb{R}. \tag{1.6}$$

From these two computational rules of the logarithm it follows (from Eq. (1.6)):

$$\log_b(x^{-1}) = -\log_b x$$

and thus from Eq. (1.5)

$$\log_b x - \log_b y = \log_b x + \log_b y^{-1} = \log_b(x \cdot y^{-1}) = \log_b\left(\frac{x}{y}\right).$$

Converting Logarithms to Different Bases

It is often practically interesting to calculate logarithms to different bases. The conversion from base a to base b is in the form:

$$\log_b y = \frac{\log_a y}{\log_a b}. \tag{1.7}$$

This results from the definition of the logarithm and the rules of calculation:

$$b^x = y \iff x = \log_b y$$

$$b^x = y \iff \log_a b^x = \log_a y \iff x \cdot \log_a b = \log_a y \iff x = \frac{\log_a y}{\log_a b}$$

and thus:

$$\boxed{\log_b y = \frac{\log_a y}{\log_a b}.} \tag{1.8}$$

Particular importance is given to logarithms with base 10 (decadic logarithm):

$$\log_{10} x = \lg x$$

and with base e (natural logarithm):

$$\log_e x = \ln x,$$

where e is the Euler number

$$e \approx 2.718\ldots. \tag{1.9}$$

1.3 Basic Arithmetic Operations

Sequences and Series

A sequence of numbers

$$a_1, a_2, a_3, \ldots, a_n, a_{n+1}, \ldots \tag{1.10}$$

is an arrangement of real numbers. The individual members of the sequence may possibly be calculated according to a rule from natural numbers. In the sequence (1.10) a_n denotes the general term. The rule for the sequence results either from the expression for a_n (explicit definition)[9] or from the connection between the sequence members. This second way of defining sequences is called **recursive definition**. If there is a natural number k and real numbers c_j, $j = 1, \ldots, k$ with the property that from a certain n on for all sequence members

$$a_{n+k} = c_1 a_{n+k-1} + c_2 a_{n+k-2} + \cdots + c_k a_k, \quad (n \geq k \geq 1) \tag{1.11}$$

applies, then the sequence (1.10) is called a *recursive sequence of k-th order* and Eq. (1.11) is called the **recursion equation**. We denote a sequence with $(a_n)_{n \geq 1}$.[10]

Examples

1. A sequence of numbers is called an **arithmetic sequence** if the difference between two successive members always has the same value. The arithmetic sequence can be described by the rule:

$$a_{n+1} = a_n + d, \quad d \text{ constant} \tag{1.12a}$$

or the explicit definition

$$a_n = a_1 + (n-1) \cdot d, \quad d \text{ constant.} \tag{1.12b}$$

Note: Eq. (1.12a) is not in recursive form, because on the right side only sequence members with constant coefficients may occur, as Eq. (1.11) shows. However, if we consider two adjacent values

$$a_{n+2} = a_{n+1} + d \quad \text{and} \quad a_{n+1} = a_n + d,$$

then by subtracting

$$a_{n+2} - a_{n+1} = a_{n+1} - a_n,$$

or

$$a_{n+2} = 2a_{n+1} - a_n. \tag{1.13}$$

---

[9] We follow the concept formation of Arens et al. (2018), Sect. 6.1.
[10] We have chosen 1 as the starting value. Note that you can start a sequence with any other integer.

Eq. (1.13) is the recurrence equation of the arithmetic sequence, so it is a recursive sequence of second order.

2. A sequence of numbers is called a **geometric sequence** if the quotient of two successive members always has the same value.

$$a_{n+1} = q \cdot a_n \qquad (1.14a)$$

or with the explicit definition

$$a_n = q^{n-1} \cdot a_1. \qquad (1.14b)$$

Eq. (1.14a) is the recurrence equation of the geometric sequence. This implies that the geometric sequence is a recursive sequence of first order.

3. The **harmonic sequence** is the sequence of numbers that results from the reciprocals of the positive integers:

$$a_n = \frac{1}{n}, \quad n = 1, 2, \ldots.$$

This results in the sequence

$$1, \frac{1}{2}, \frac{1}{3}, \frac{1}{4}, \ldots$$

The harmonic sequence is not recursive.

4. The **Fibonacci sequence** is recursively defined by[11]

$$F_{n+2} = F_{n+1} + F_n,$$

with the two initial values $F_0 = F_1 = 1$. The first terms of the sequence are:

$$1, 1, 2, 3, 5, 8, 13, 21, \ldots$$

The Fibonacci sequence is a recursive sequence of second order. The explicit definition of the members of the sequence F_n as a function of the number n is the Binet formula:

$$F_n = \frac{1}{\sqrt{5}} \left\{ \left(\frac{1+\sqrt{5}}{2} \right)^n - \left(\frac{1-\sqrt{5}}{2} \right)^n \right\}.$$

5. An important sequence that is related to Euler's number (1.9) is

$$a_n = \left(1 + \frac{1}{n}\right)^n, \quad n = 1, 2, 3, \ldots. \qquad (1.15)$$

---

[11] The Fibonacci sequence has very interesting and far-reaching properties. We refer to the literature for this, see Koshy (2001) or Posamentier and Lehmann (2007).

1.3 Basic Arithmetic Operations

The first terms of this sequence are

$$a_1 = 2, \; a_2 = 2,25, \; a_3 = 2,37, \; a_4 = 2,44,\ldots$$

If the terms $a_1, a_2,\ldots, a_n$ of a finite sequence are added, a **finite series** is obtained.

$$a_1 + a_2 + \cdots + a_n = \sum_{i=1}^{n} a_i = s_n. \tag{1.16}$$

The sums over finitely many sequence terms s_n are called partial sums.

Examples

1. The partial sums of the arithmetic sequence are:

$$s_n = \sum_{i=1}^{n} a_i = \frac{n}{2}(a_1 + a_n). \tag{1.17}$$

2. The finite geometric series is:

$$\begin{aligned} s_n &= a_1 + a_2 + \cdots + a_n \\ &= a_1 + a_1 q + a_1 q^2 + \cdots + a_1 q^{n-1} \\ &= a_1 \left(1 + q + q^2 + \cdots + q^{n-1}\right) \\ &= a_1 \cdot \sum_{i=1}^{n} q^{i-1} \\ &= a_1 \cdot \frac{q^n - 1}{q - 1}. \end{aligned} \tag{1.18}$$

Eq. (1.18) applies to all $q \in \mathbb{R}$ which are different from 0 and 1.[12]

3. The sequence of the first n natural numbers is

$$1, 2, 3, \ldots n;$$

this leads to the partial sums

$$s_n = 1 + 2 + 3 + \cdots + n = \sum_{i=1}^{n} i = \frac{n(n+1)}{2}.$$

This form of summation of the natural numbers goes back to Gauss.[13]

---

[12] See also the derivation of Eq. (2.12) in Sect. 2.6.1.

[13] One shows such relations, which apply to all natural numbers, by means of an important proof technique, which one calls complete induction. See for example Lang (1986, p. 87), or Spivak (2008), Chap. 2.

Binomial Formulas

The generalization of the binomial formulas:

$$(a+b)^2 = a^2 + 2ab + b^2$$
$$(a-b)^2 = a^2 - 2ab + b^2$$
$$(a+b)(a-b) = a^2 - b^2$$

is the binomial expression (Binomial Theorem)

$$(a+b)^n = \sum_{k=0}^{n} \binom{n}{k} a^{n-k} b^k \qquad (1.19)$$

with the summation

$$\sum_{k=0}^{n} x_k = x_0 + x_1 + \ldots + x_n$$

and the **Binomial Coefficients**

$$\binom{n}{k} = \frac{n!}{(n-k)!k!} \qquad \text{(read: } n \text{ choose } k\text{)}, \qquad (1.20)$$

where

$$n! = 1 \cdot 2 \cdot 3 \cdot \ldots \cdot (n-1) \cdot n \qquad \text{(read: } n \text{ factorial)}$$

and we define $0! = 1$.

The binomial coefficients satisfy the relation:

$$\binom{n}{k} = \binom{n-1}{k-1} + \binom{n-1}{k}.$$

This relation is known as the **Pascal's Triangle**[14]

1.4 Equations

In this section we consider a number of equations that can be solved by equivalent transformations using the basic arithmetic operations from Section 1.3.

Under the **solution set** of an equation $A(x) = B(x)$ we understand the set

$$L = \{x \mid A(x) = B(x) \text{ is a true statement}\}.$$

---

[14] Further discussion of the Binomial coefficients can be found in Graham et al. (1994).

1.4 Equations

The solution set is a subset of the definition set that specifies for which $x \in \mathbb{R}$ $A(x)$ and $B(x)$ are defined.

Under an **equivalent transformation** of an equation, one understands, following equivalent proposition forms (cf. section 1.1), such transformations that do not change the solution set. Under an equivalent transformation of an equation, we denote transformations that do not change the set of solutions. This is based on equivalent proposition forms (cf. Section 1.1):

$$A(x) = B(x) \iff \widetilde{A(x)} = \widetilde{B(x)}$$

if

$$L = \{x \mid A(x) = B(x) \text{ is a true statement }\}$$
$$= \{x \mid \widetilde{A(x)} = \widetilde{B(x)} \text{ is a true statement }\}.$$

Equivalent transformations are:

1. Addition of a term $T(x)$

$$A(x) = B(x) \qquad \Big| + T(x)$$
$$\iff A(x) + T(x) = B(x) + T(x).$$

Example

$$3x + 5 = 2x - 1 \qquad \Big| - 2x - 5$$
$$\iff \quad x = -6.$$

2. Multiplication with a term $T(x)$

$$A(x) = B(x) \qquad \Big| \cdot T(x), T(x) \neq 0$$
$$\iff T(x) \cdot A(x) = T(x) \cdot B(x).$$

Note: For $T(x) = 0$ there is no equivalence transformation.

Example

$$\frac{1}{x-1} = \frac{2}{x-7} \qquad \Big| \cdot (x-1)(x-7), x \neq 1 \land x \neq 7$$
$$\iff x - 7 = 2(x-1)$$
$$\iff \quad x = -5.$$

3. Logarithm to a base b

$$A(x) = B(x) \quad | \log_b \text{ for } A(x), B(x) > 0$$
$$\iff \log_b A(x) = \log_b B(x).$$

Example

$$10^x = 25 \quad | \lg$$
$$\iff x = \lg 25 \approx 1.398.$$

4. Raising to a power b

$$A(x) = B(x) \quad | b^{(\ )} \text{ for } b \neq 1$$
$$\iff b^{A(x)} = b^{B(x)}.$$

Example

$$\log_5(3x) = 2 \quad | 5^{(\ )}, x > 0$$
$$\iff 5^{\log_5(3x)} = 5^2$$
$$\iff 3x = 25$$
$$\iff x = \frac{25}{3}.$$

5. When raising an equation to a power, it must be distinguished whether n is even or odd:

$$A(x) = B(x) \quad | (\ )^n \text{ for } n \text{ odd}$$
$$\iff (A(x))^n = (B(x))^n,$$

and

$$A(x) = B(x) \quad | (\ )^{\frac{1}{n}} \text{ for } n \text{ odd}$$
$$\iff (A(x))^{\frac{1}{n}} = (B(x))^{\frac{1}{n}}.$$

Raising to a power for even n is not an equivalent transformation.

With these equivalent transformations, we now solve the **linear** and the **quadratic equation** in full generality.

The Linear Equation
When solving the linear equation

$$ax + b = 0$$

1.4 Equations

three cases are distinguished:

Case 1: $a \in \mathbb{R}, a \neq 0$ and $b \in \mathbb{R}$, then the unique solution is

$$x = -\frac{b}{a}.$$

Case 2: $a = 0 \land b \neq 0$, then there is no solution.

Case 3: $a = 0 \land b = 0$, then any number $x \in \mathbb{R}$ is a solution of the linear equation (arbitrarily many solutions).

The quadratic equation

$$ax^2 + bx + c = 0 \qquad a \neq 0$$
$$\iff x^2 + \frac{b}{a}x + \frac{c}{a} = 0$$
$$\iff \left(x + \frac{b}{2a}\right)^2 - \frac{b^2}{4a^2} + \frac{c}{a} = 0$$
$$\iff \left(x + \frac{b}{2a}\right)^2 = \frac{b^2}{4a^2} + \frac{c}{a}$$
$$\iff x_{1/2} = \frac{-b \pm \sqrt{b^2 - 4ac}}{2a}.$$

Remark For $b/a = p$ and $c/a = q$ it follows:

$$x_{1/2} = -\frac{p}{2} \pm \sqrt{\left(\frac{p}{2}\right)^2 - q}.$$

The quadratic equation has:

- two real solutions for $b^2 - 4ac > 0$,
- one real solution for $b^2 - 4ac = 0$,
- no real solution for $b^2 - 4ac < 0$.

Equivalent transformations of inequalities

When equivalent transformations of **inequalities** are considered, the following rules must be observed:

1. Addition with a constant c:

$$A(x) < B(x)$$
$$\iff A(x) + c < B(x) + c.$$

$$A(x) > B(x)$$
$$\iff A(x) + c > B(x) + c.$$

2. Multiplication by a constant $c \neq 0$:

$$A(x) < B(x)$$
$$\iff cA(x) < cB(x) \quad \text{if } c > 0$$
$$\iff cA(x) > cB(x) \quad \text{if } c < 0.$$

$$A(x) > B(x)$$
$$\iff cA(x) > cB(x) \quad \text{if } c > 0$$
$$\iff cA(x) < cB(x) \quad \text{if } c < 0.$$

3. Power:

$$A(x) < B(x)$$
$$\iff (A(x))^n < (B(x))^n.$$

$$A(x) > B(x)$$
$$\iff (A(x))^n > (B(x))^n.$$

4. Root:

$$A(x) < B(x)$$
$$\iff (A(x))^{\frac{1}{n}} < (B(x))^{\frac{1}{n}}.$$

$$A(x) > B(x)$$
$$\iff (A(x))^{\frac{1}{n}} > (B(x))^{\frac{1}{n}}.$$

5. Logarithm:

$$A(x) < B(x)$$
$$\iff \log_b(A(x)) < \log_b(B(x)), \text{ for } b > 1.$$

$$A(x) > B(x)$$
$$\iff \log_b(A(x)) > \log_b(B(x)), \text{ for } b > 1.$$

6. Power to a base a:

$$A(x) < B(x)$$
$$\iff a^{A(x)} < a^{B(x)}, \text{ for } a > 1.$$

$$A(x) > B(x)$$
$$\iff a^{A(x)} > a^{B(x)}, \text{ for } a > 1.$$

and
$$A(x) < B(x) \iff a^{-A(x)} > a^{-B(x)}, \text{ for } a > 1.$$

$$A(x) > B(x) \iff a^{-A(x)} < a^{-B(x)}, \text{ for } a > 1.$$

Equations with absolute values

The **absolute value** of a number $a \in \mathbb{R}$ is defined by:

$$|a| = \begin{cases} a & \text{for } a \geq 0 \\ -a & \text{for } a < 0. \end{cases}$$

Therefore, calculating with absolute values leads to case distinctions.
  The solution of the equation

$$|ax + b| = c \quad \text{where } c \geq 0$$

leads to:

$$ax + b = c \quad \vee \quad -(ax + b) = c$$

with the solutions

$$x = \frac{c - b}{a} \quad \vee \quad x = -\frac{c + b}{a}.$$

1.5 Trigonometry

The trigonometric functions are based on the elementary geometry of right-angled triangles (Fig. 1.8).

$$\sin \alpha = \frac{b}{c} = \frac{\text{opposite side}}{\text{hypothenuse}}$$

$$\cos \alpha = \frac{a}{c} = \frac{\text{adjacent side}}{\text{hypothenuse}}$$

$$\tan \alpha = \frac{b}{a} = \frac{\text{opposite side}}{\text{adjacent side}}$$

$$\cot \alpha = \frac{a}{b} = \frac{\text{adjacent side}}{\text{opposite side}}.$$

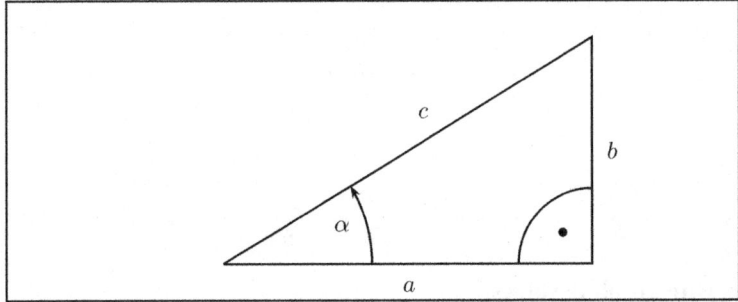

Fig. 1.8 To define the trigonometric functions by looking at the side ratios in a right-angled triangle

In analysis, the tangent plays an important role, as this function is used to calculate the slope of tangents.

This book does not contain any exercises on the elementary basics. We refer to a variety of books on this subject, including: Bosch (2010), S. Lang (1988), Purkert (2014) or Schäfer et al. (2006).

1.6   Test

1.1 Form the sets $A \cup B$, $A \cap B$ and $A \smallsetminus B$:

$$A = \{x \mid x \in \mathbb{N} \land x \leq 10\}$$
$$B = \left\{x \mid x \leq 20 \land \frac{x}{2} \in \mathbb{N}\right\}.$$

1.2 Are the following statement forms $A(x)$ and $B(x)$ equivalent?

(a) $A(x): x = \pm 4$; $B(x): x^2 = 16$.
(b) $A(x): x^2 - y^2 = 0$; $B(x): x = y \lor x = -y$.
(c) $A(x): x^2 \geq a$; $B(x): x \geq \sqrt{a}$.

1.3

(a) Name a sufficient and a necessary condition for winning the lottery.
(b) Is $A(x)$ a sufficient condition for $B(x)$?

$$A(x): x > 0; \quad B(x): \log_2 x > 0.$$

1.6 Test

1.4 Simplify the following terms, for which x the terms are not defined?

(a) $\frac{x+2}{x^2-4} + \frac{1}{x+2}$

(b) $\frac{3x^n + 2x^{n+2}}{x^{n+1} + 3x^n}$

(c) $\sqrt[3]{x^{6n-9}}$; $n \in \mathbb{N}$

1.5 Solve the following equations for x:

(a) $2x - 7 = \frac{3}{2}x + \sqrt{3}$

(b) $\frac{x-3}{2x+6} = 4$

(c) $3x^2 + 2x - 1 = 0$

(d) $x - 3 = \frac{1}{2+x}$

(e) $\sqrt{x} = 1 - x$.

(f) $x^5 - 12 = 3$

(g) $x^4 + 4x^2 - 8 = 0$

(h) $3^x + 12 = 24$

(i) $3^x + 3^{x+2} = 110$

(j) $-2^x + 4 \cdot 2^{2x} = 128$

(k) $\log_2 4x = 15$

(l) $\lg(x + 10) - \lg(2x + 5) = 3$

(m) $\log_2 3x + \log_4 5x = 3$.

1.6 Determine the positive real number x with the property that the number $1/x$ is 1 less than x. What property does the number x^2 have?

1.7 Calculate $(a + b)^4$.

1.8 Illustrate the relationship

$$\binom{n}{k} = \binom{n-1}{k-1} + \binom{n-1}{k},$$

and show formally its validity.

1.9 Determine the solution set of the following equations:

(a) $3x + 5 \geq -2x - 3$

(b) $x^5 > 125$

(c) $-2x^2 + 3x - 1 < 0$.

1.10 Show that the solution of the quadratic equation

$$x^2 - px + q = 0$$

is equivalent to finding two numbers x_1, x_2 with

$$x_1 + x_2 = p$$
$$x_1 \cdot x_2 = q.$$

The solutions to this test can be found after Chap. 7.

References

Arens T., Hettlich F., Karpfinger Ch., Kockelkorn U., Lichtenegger K., Stachel H. (2018): Mathematik, 4. Auflage, Spektrum Akademischer Verlag, Heidelberg.
Basieux P. (2000): Die Architektur der Mathematik, Denken in Strukturen, Rowohlt, Hamburg.
Bosch K. (2010): Brückenkurs Mathematik, Eine Einführung mit Beispielen und Übungsaufgaben, 14. Auflage, Oldenbourg, München.
Dean N. (2003): Diskrete Mathematik, Pearson Education, München.
Deiser O. (2010): Einführung in die Mengenlehre, 3. Edition, Springer, Berlin, Heidelberg.
Garnier R., Taylor J. (2002): Discrete Mathematics for New Technology, Bristol, Philadelphia.
Graham R.L., Knuth D.E., Patashnik O. (1994), Concrete Mathematics, Second Edition, Addison-Wesley, Boston.
Gregg J. R. (1998): Ones and Zeros, Understanding Boolean Algebra, Digital Circuits and the Logic of Sets, IEEE Press, New York.
Hilgert I., Hilgert J. (2021): Mathematik – ein Reiseführer, 2. Edition, Springer Spektrum, Heidelberg.
Kelly, J. (2003): Logik im Klartext, Pearson Studium, München.
Körner T.W. (2020): Where do numbers come from? Cambridge University Press, Cambridge UK.
Koshy T. (2001): Fibonacci and Lucas Numbers with Applications. John Wiley and Sons.
Koshy T. (2004): Discrete Mathematics with Applications, Elsevier, Amsterdam.
Kramer J., von Pippich A.-M. (2013): Von den natürlichen Zahlen zu den Quaternionen; Basiswissen Zahlbereiche und Algebra, Springer Spektrum, Wiesbaden.
Kreuzer M., Kühling S. (2006): Logik für Informatiker, Pearson Studium, München.
Lang S. (1986): A First Course in Calculus, Fifth Edition, Springer Verlag, New York.
Lang S. (1988): Basic Mathematics, Springer Verlag, New York.
Maor E. (2015): e: The Story of a Number, Princeton University Press, Princeton, New Jersey.
Purkert W. (2014): Brückenkurs Mathematik für Wirtschaftswissenschaftler, 8. aktualisierte Auflage, Springer-Gabler.
Posamentier A. S., Lehmann I. (2007): The (Fabulous) Fibonacci Numbers, Prometheus Books, New York.
Schäfer W., Georgi K., Otto Ch., Trippler G. (2006): Mathematik-Vorkurs, Übungs- und Arbeitsbuch für Studienanfänger, 6. Auflage, Vieweg Teubner, Stuttgart.
Spivak M. (2008): Calculus, Third Edition, Cambridge University Press, Cambridge.
Staab F. (2012): Logik und Algebra, 2. Auflage Oldenbourg.
Toenniessen F. (2019): Das Geheimnis der transzendenten Zahlen. 2. Auflage, Spektrum Akademischer Verlag, Heidelberg.
Winter R. (2001): Grundlagen der formalen Logik, 2. überarbeitete Auflage, Verlag Harri Deutsch, Frankfurt.

Functions   2

Learning Objectives (This chapter provides)

- How the dependence of quantitative variables is described with functions
- The required basic knowledge of elementary functions
- Basic properties of functions
- The application of functions in the context of economic questions
- The introduction of the concept of limit ◄

2.1 Definition and Representation of Functions

Both in mathematics and in many areas of everyday life, numbers or quantitatively measurable quantities are often related to each other in such a way that a dependency between such quantities arises. Such relationships are established in different ways:

- by experimental or heuristic methods
- by theoretical models
- or simply by arbitrary decision.

The following examples serve on the one hand to illustrate these considerations, on the other hand they motivate the introduction of the concept of functions.

Examples

1. The mapping of a set of students to their math grades. Let's consider a particular course with n students and the list of grades from the math final. We'll label the students as $S_1, S_2, \ldots, S_n$. Each student can earn a grade between 1.0 and 5.0, with tenths of a point being common. For example, the following final results might occur:

Student	S_1	S_2	S_3	...	S_{n-1}	S_n
Grade	2,2	2,7	3,5	...	2,2	1,6

   The mapping of student to grade can be represented as a set of ordered pairs:

   $$(S_1; 2.2), (S_2; 2.7), (S_3; 3.5), \ldots (S_{n-1}; 2.2), (S_n; 1.6).$$

   This mapping reveals the following facts, which are characteristic of the concept of a function:

   - *Each* student has *exactly* one grade.
   - In general, it is possible that some students have the same grade (e.g., 2.2).
   - It is not necessary for all grades in the possible range of grades to be achieved. That is, a final result may have no grades of 4.5 or 5.0.

2. The volume of a sphere as a function of the radius.
   As is known from geometry, the volume of a sphere V_s depends on the radius r according to

   $$V_s = \frac{4}{3}\pi r^3.$$

   Here, every value $r \geq 0$ is assigned exactly one value V_s.

With these preliminary considerations, we consider the definition of a function.

▶ **Definition (Function)** Let M_1 and M_2 be two sets. If one assigns *each* element of the set M_1 uniquely *exactly* one element of the set M_2, then the resulting assignment is called a **function**.

According to this definition, there must be no element left in M_1, and the mapping must be *unique*. This means *not* that each element from M_1 is associated with a different element from M_2, but that one element from M_1 is not associated with two or more elements from M_2. Fig. 2.1 shows the assignment Student → Mark in the form of a diagram. The set M_1 is formed by the students, i.e.:

$$M_1 = \{S_1, S_2, \ldots, S_{n-1}, S_n\}$$

2.1 Definition and Representation of Functions

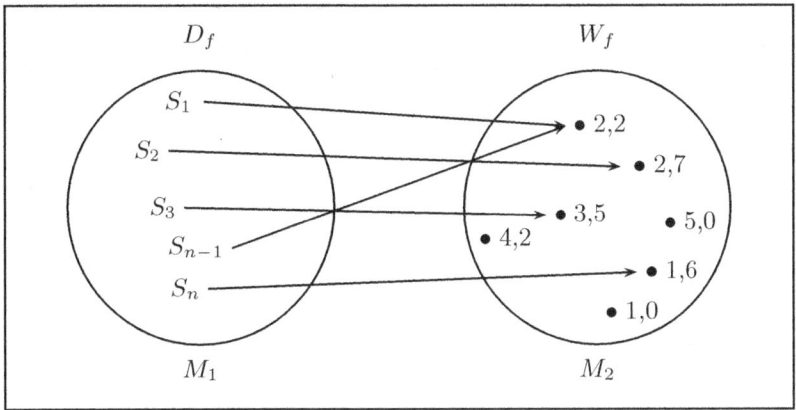

Fig. 2.1 Representation of the function Student → Mark in the form of a diagram. Here, the set of students of a course $M_1 = D_f$ and the set of grades is $W_f \subset M_2$

and the set M_2 are the achievable grades:

$$M_2 = \{1.0;\ 1.1;\ 1.2;\ \ldots;\ 4.9;\ 5.0\}.$$

The following terms are important in connection with the definition of a function:

- The 'left' set M_1 from Fig. 2.1, from which the mapping arrows originate, is called the **domain** or **range** D_f of the function f.
- The 'right' set M_2, in which the mapping arrows end, is called the **codomain** or **image**.
- The subset of M_2 whose elements are actually mapped to a domain element is called the **range** W_f of the function f.

Functions can be represented in different forms:

1. In the form of a diagram, as in Fig. 2.1.
2. As a set of ordered pairs:

$$f = \{(S_1;\ 2.2),\ (S_2;\ 2.7),\ (S_3;\ 3.5),\ \ldots\ (S_{n-1};\ 2.2),\ (S_n;\ 1.6)\}$$

as in Example 1.
3. Functions over the real numbers $\mathbb{R}$ are often represented in the form of a mapping:

$$f : D_f \longrightarrow W_f,$$
$$x \longmapsto y = f(x).$$

The *maximum allowable domain* includes all possible values x for which $f(x)$ is defined.

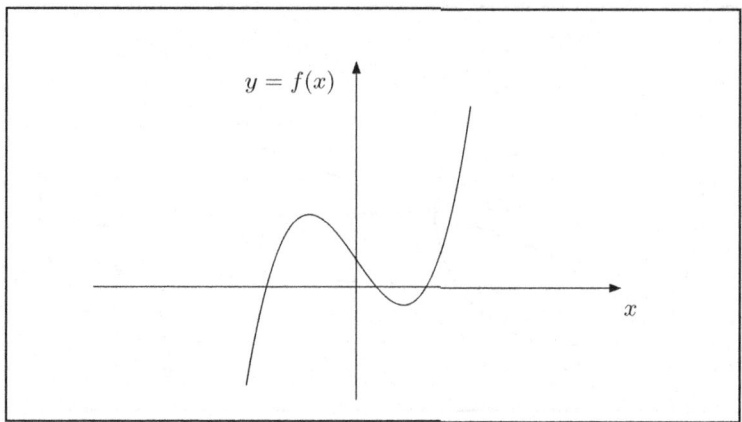

Fig. 2.2 Representation of a function in the coordinate system

4. As a representation in the coordinate system:
   The independent variable is represented on the *x*-axis (abscissa), the dependent variable on the *y*-axis (ordinate) (Fig. 2.2).

2.2 Some Elementary Functions

In this section we will take a closer look at a number of elementary functions and recapitulate the essential properties.

2.2.1 Linear Function

The linear function is a mapping of the form

$$f : \mathbb{R} \longrightarrow \mathbb{R},$$
$$f : x \longmapsto f(x) = ax + b$$

with the two parameters $a, b \in \mathbb{R}$, which determine the shape of the line. a determines the slope of the line, the parameter b the distance to the *x*-axis at $x = 0$. The graph of this function is therefore a line through the point $x = 0, y = b$, which in the case $a \neq 0$ also goes through the point $x = -\frac{b}{a}, y = 0$. In the case $a = 0$, the line runs parallel to the *x*-axis. For $a \neq 0$, the slope triangle of the line is obtained by going from any point of the line one unit to the right and a units in *y*-direction.

In the case $b = 0$, one obtains a line that runs through the origin of the coordinate system. In this case one says that y is *proportional* to x (Fig. 2.3).

2.2 Some Elementary Functions

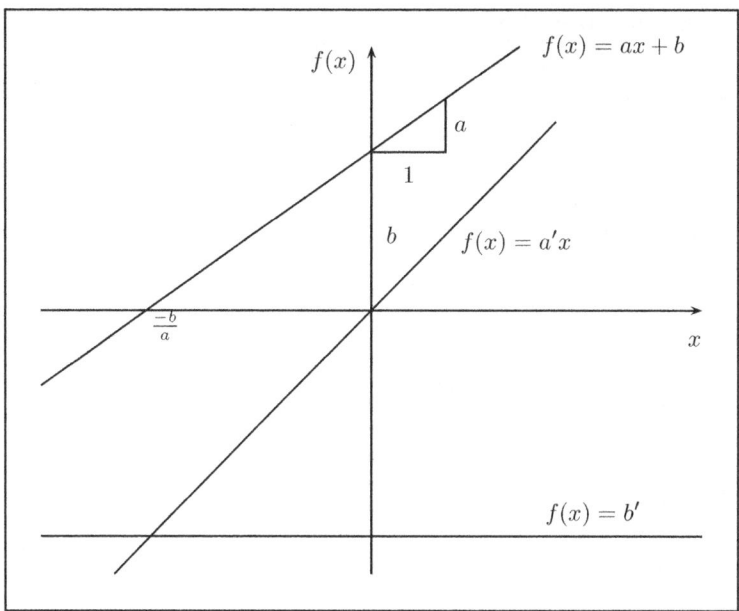

Fig. 2.3 Representation of linear functions in the coordinate system

In the case $a = 0$, $b = b'$ one obtains the function:

$$f(x) = b'.$$

It is also referred to as the *constant function*.

Notes

1. Lines of the form $x = c = const.$ cannot be represented by the linear equation

$$y = ax + b.$$

They characterize parallels to the *y*-axis and are not functions, since the uniqueness of the assignment is not given.

2. If a line is determined by the slope m and a point $P = (x_1, y_1)$, it is advisable to use the **point-slope form of the linear equation**:

$$y = m \cdot (x - x_1) + y_1. \tag{2.1}$$

3. If a line is determined by two points $P_1 = (x_1, y_1)$ and $P_2 = (x_2, y_2)$, then the **two-point form of the line equation is useful**

$$\frac{y - y_1}{x - x_1} = \frac{y_2 - y_1}{x_2 - x_1}. \tag{2.2}$$

2.2.2 Quadratic Function

The quadratic function is also called *Parabola* The most general form of this function is:
$$f : \mathbb{R} \longrightarrow \mathbb{R},$$
$$f : x \longmapsto f(x) = ax^2 + bx + c$$
with three real parameters $a, b, c \in \mathbb{R}$, which determine the shape of the parabola. The graph of this function is shown in Fig. 2.4.

The zeros of this function—these are the points at which the graph intersects the x-axis—are given by

$$\boxed{x_{1/2} = \frac{-b \pm \sqrt{b^2 - 4ac}}{2a}}$$

From this one can see the following situation:

- If $b^2 > 4ac$, then there are two real solutions, that means the graph has two intersections with the x-axis.
- If $b^2 = 4ac$, then the parabola has exactly one zero, the graph of the function intersects the x-axis in this case only once.
- If $b^2 < 4ac$, then the parabola has no zeros, the graph doesn't intersect the x-axis.

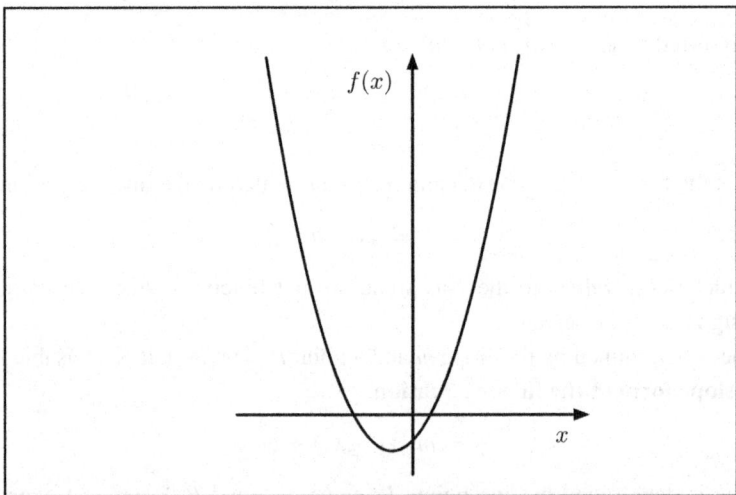

Fig. 2.4 Graph of the quadratic function $f(x) = ax^2 + bx + c$

2.2 Some Elementary Functions

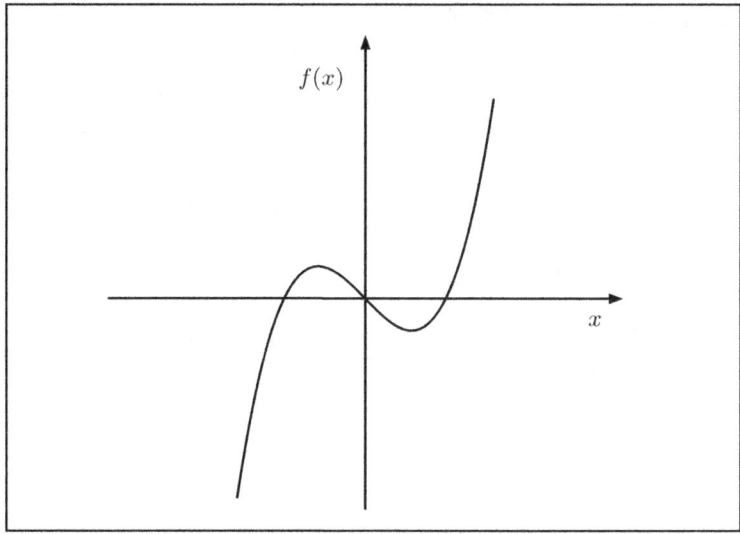

Fig. 2.5 The graph of the polynomial function of 3rd degree $f(x) = x^3 - x$

2.2.3 Integer Rational Functions or Polynomials

The linear and quadratic functions already introduced above are special cases of a much larger class of functions, the so-called *polynomia functions*. These have the general form:

$$f : \mathbb{R} \longrightarrow \mathbb{R}$$

with:

$$f(x) = a_n x^n + a_{n-1} x^{n-1} + \ldots + a_2 x^2 + a_1 x^1 + a_0$$
$$= \sum_{i=0}^{n} a_i x^i.$$

The fixed parameters $a_0, a_1, \ldots a_n$ are called coefficients and are real numbers. In Fig. 2.5 the graph of a whole polynomial function for $a_3 = 1$, $a_1 = -1$, $a_2 = a_0 = 0$ is shown.

A polynomial function, whose coefficients a_n of the highest power x^n are not equal to zero, is also called **polynomial of degree** n.

2.2.4 Power Function

Power functions are whole rational functions with

$$f(x) = x^n; \; n \in \mathbb{N}_0.$$

They are also referred to as **parabolas** of n-th order.

2.2.5 Fractional Rational Functions

If you form the quotient of two polynomial functions, you get the general form of the *fractional rational functions*:

$$f(x) = \frac{a_n x^n + a_{n-1} x^{n-1} + \ldots + a_2 x^2 + a_1 x^1 + a_0}{b_m x^m + b_{m-1} x^{m-1} + \ldots + b_2 x^2 + b_1 x^1 + b_0} \qquad (2.3)$$

with $m, n \in \mathbb{N}$. These functions are defined everywhere where the denominator is not equal to zero.

For some investigations, a representation of a fractional rational function as a sum is helpful.[1] This is achieved by the so-called polynomial division. For this, the fractional rational function

$$f(x) = \frac{P(x)}{Q(x)},$$

where $P(x)$ is a polynomial function of degree m and $Q(x)$ is a polynomial function of degree n for n > m is represented in the form

$$f(x) = P_1(x) + \frac{P_2(x)}{Q(x)}.$$

In this case, $P_1(x)$ and P_2 are polynomial functions, where P_2 has a smaller degree than $Q(x)$. The execution of the polynomial division corresponds to the written division and is exemplarily easily understandable.

Example Consider

$$f(x) = \frac{6x^3 + 5x^2 - 3x + 1}{3x - 2} = \frac{P(x)}{Q(x)},$$

where the degree of $P(x)$ is n =3 and that of $Q(x)$ is m =1. The polynomial division is then:

$$
\begin{array}{l}
(6x^3 + 5x^2 - 3x + 1) : (3x - 2) = 2x^2 + 3x + 1 + \dfrac{3}{3x - 2} \\
\underline{6x^3 - 4x^2} \\
\qquad 9x^2 - 3x + 1 \\
\qquad \underline{9x^2 - 6x} \\
\qquad\qquad 3x + 1 \\
\qquad\qquad \underline{3x - 2} \\
\qquad\qquad\qquad 3
\end{array}
$$

---
[1] See, for example, Sect. 2.6.6.

Therefore, it follows:

$$f(x) = \frac{6x^3 + 5x^2 - 3x + 1}{3x - 2} = 2x^2 + 3x + 1 + \frac{3}{3x - 2}$$

with polynomial functions

$$P_1(x) = 2x^2 + 3x + 1, \quad P_2(x) = 3, \quad Q(x) = 3x - 2.$$

The polynomial division is also used in the determination of zeros of polynomials of degree >2, if at least one zero is already known.[2]

2.2.6 Hyperbolic Function

The simplest fractional rational function is the Hyperbolic function, it is a mapping of the form:

$$f : \mathbb{R} \setminus \{0\} \longrightarrow \mathbb{R},$$
$$f : x \longmapsto f(x) = \frac{a}{x}; \quad a \in \mathbb{R}.$$

You can find the following situation by creating a value table for this function: If the x values tend to zero for positive real numbers, the function values a/x grow over all (positive) limits. If the x values tend to zero for negative real numbers, the function values grow over all (negative) limits.

The image of this function (see Fig. 2.6) is a curve lying symmetrically to the angle bisectors of the axes, a hyperbola. This function is not defined for the point $x = 0$, since the division by zero is not defined.

The connection between $f(x)$ and x is also referred to as inversely proportional.

2.2.7 Root Function

The **root function** is the inverse function (cf. Sect. 2.3) of the power function. It is defined by:

$$f : \mathbb{R}_0^+ \longrightarrow \mathbb{R}_0^+,$$
$$f : x \longmapsto f(x) = x^{\frac{1}{n}} = \sqrt[n]{x}; \quad x \in \mathbb{R}_0^+, \quad n \in \mathbb{N}.$$

---

[2] See Sect. 3.6.2.

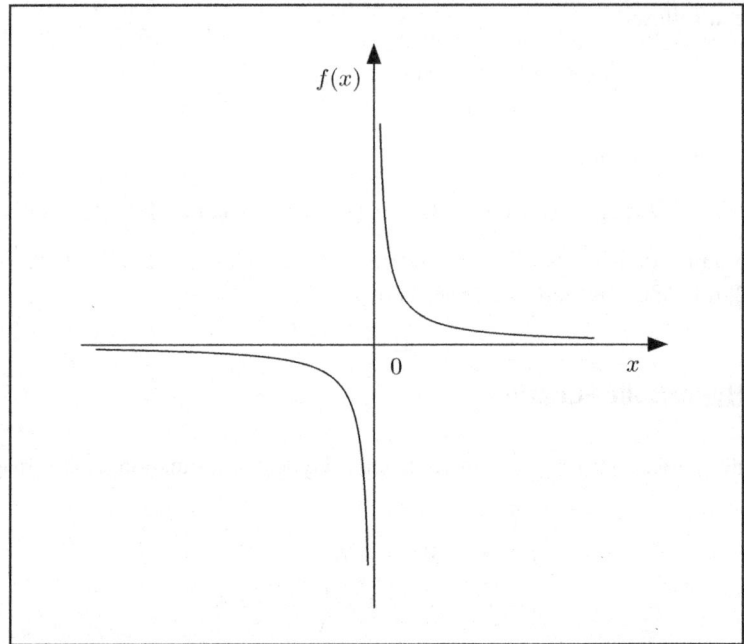

Fig. 2.6 The function $f(x) = 1/x$

Note

For odd $n \in \mathbb{N}$, the domain can be extended to $\mathbb{R}$.

For $n = 2$, we get:
$$f(x) = \sqrt{x} = x^{\frac{1}{2}}.$$
The graph of the root function $f(x) = \sqrt{x}$ is shown in Fig. 2.7.

2.2.8 Exponential Function

The **exponential function** is defined by the mapping rule:

$$f : \mathbb{R} \longrightarrow \mathbb{R}^+,$$
$$f : x \longmapsto f(x) = a^x, \; x \in \mathbb{R}; \; a > 0; \; a \neq 1.$$

The parameter a is called the **base** of the exponential function. In particular, the exponential function with base $a = e \approx 2.71828$ plays a major role.[3] It is also briefly referred to as the e-function. We occasionally use the notation $e^x = \exp(x)$.

---

[3] See the recommended book by Maor (2015) on the history of the development of the number e.

2.2 Some Elementary Functions

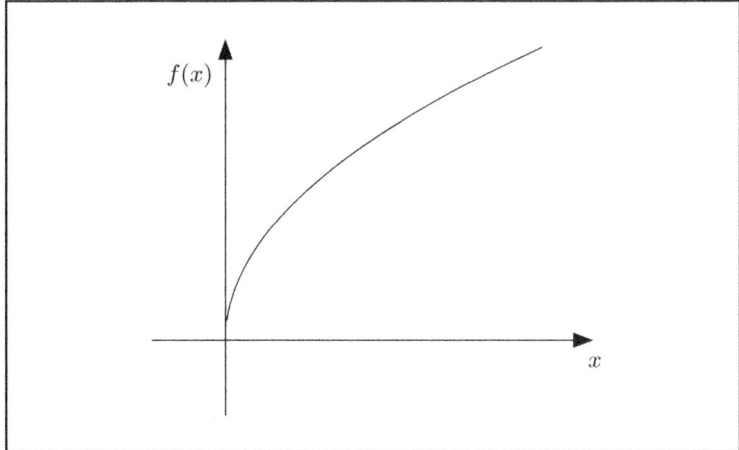

Fig. 2.7 The root function $f(x) = \sqrt{x}$

Notes

1. The exponential function has the property
$$f(0) = a^0 = 1 \text{ for all } a > 0.$$

2. Every exponential function can be transformed to the base e. Since
$$a = e^{\ln a},$$
one gets :
$$a^x = (e^{\ln a})^x = e^{x \cdot \ln a}.$$

3. The exponential functions play a major role in the applications—both in the description of technical and economic processes—because with these functions decay—or growth processes can be modeled.

In the discussion of the properties of the exponential function $f(x) = a^x$ one distinguishes two cases:

- **Case 1:** $a > 1$
  As can be seen from Fig. 2.8, the function values increase for increasing x values over all limits. The function value $f(x)$ approaches 0 when x takes on larger and larger negative values.
- **Case 2:** $0 < a < 1$
  The function values approach to zero for large positive values of x, they increase to limit for larger and larger negative values of x.

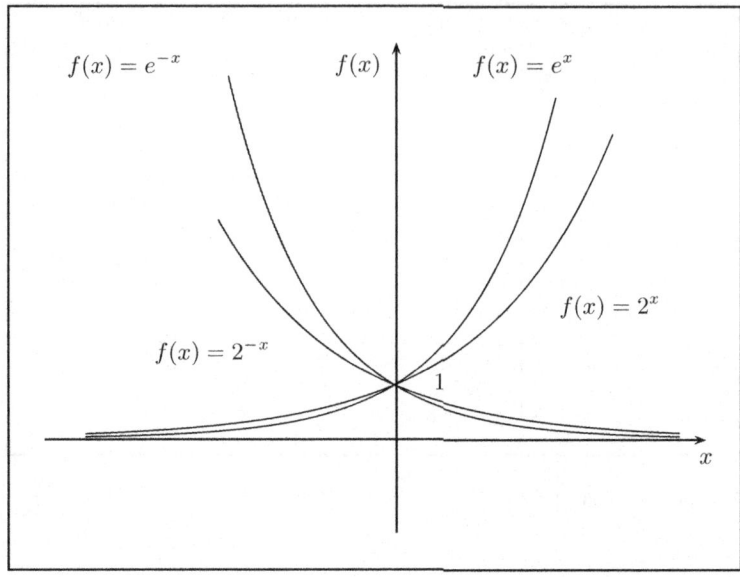

Fig. 2.8 The exponential functions $f(x) = e^x$, $f(x) = 2^x$, $f(x) = e^{-x}$ and $f(x) = 2^{-x}$

2.2.9 Logarithm Function

The logarithm function is the inverse function (see Sect. 2.3) of the exponential function.

$$f : \mathbb{R}^+ \longrightarrow \mathbb{R},$$
$$f : x \longmapsto f(x) = \log_a x, \quad a > 0,\ a \neq 1.$$

The parameter a denotes the base of the logarithm function.
Frequently used bases are:

- The natural logarithm function with base e

$$f(x) = \ln x.$$

- The decimal logarithm function with base 10

$$f(x) = \log_{10} x = \lg x.$$

Properties:

- As Fig. 2.9 shows, the value of the logarithm function increases with increasing x values.
- The logarithm function is only defined for positive real values.

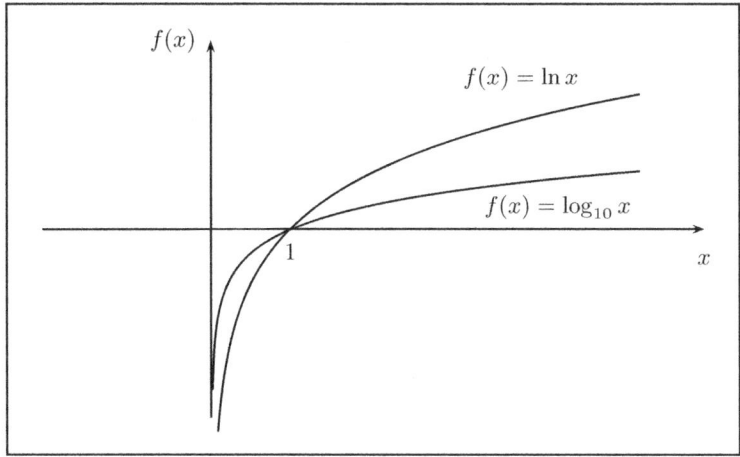

Fig. 2.9 Graph of the two logarithm functions $f(x) = \ln x$ and $f(x) = \log_{10} x$

- The graph of the logarithm function intersects the x-axis at exactly one point, at $x = 1$. This means that the logarithm function has—for each base a—the zero:
$$\log_a(1) = 0.$$

- As the x-values approach 0 ($x > 0$), the values of the logarithmic function take on arbitrarily large negative values.

2.2.10 Trigonometric Functions

The two trigonometric functions
$$f(x) = \sin x \quad \text{and} \quad g(x) = \cos x$$
are mappings with the domain
$$D_{\sin} = D_{\cos} = \mathbb{R}.$$
Sine and cosine are defined for all $x \in \mathbb{R}$ and have the range:
$$W_{\sin} = W_{\cos} = [-1, +1] = \{x \in \mathbb{R} \mid -1 \leq x \leq +1\}.$$
These trigonometric functions arise from an extension of the sine and cosine definition given in Chap. 1, which can be represented in the unit circle as shown in Fig. 2.10.

The two functions satisfy the periodicity:
$$\sin(x + n2\pi) = \sin x, \quad n \in \mathbb{Z},$$
$$\cos(x + n2\pi) = \cos x, \quad n \in \mathbb{Z}.$$

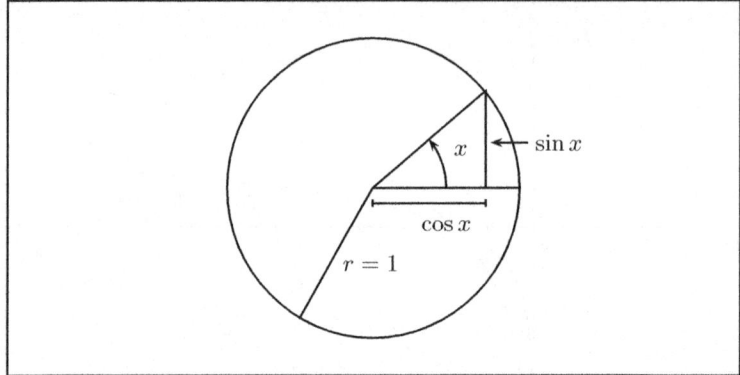

Fig. 2.10 Representation of the sine and cosine function in the unit circle

The two functions have two zeros in the interval

$$I = \{0 \leq x < 2\pi\}$$

because:

$$\sin 0 = \sin \pi = 0$$

$$\cos \frac{\pi}{2} = \cos \frac{3\pi}{2} = 0.$$

Due to the periodicity, sine and cosine functions have an infinite number of zeros $x_n, n \in \mathbb{Z}$ in the entire domain of definition $\mathbb{R}$, which are given by

$$\sin x_n = 0 \quad \Longleftrightarrow \quad x_n = n \cdot \pi, \quad n \in \mathbb{Z}$$

$$\cos x_n = 0 \quad \Longleftrightarrow \quad x_n = (2n+1) \cdot \frac{\pi}{2}, \quad n \in \mathbb{Z}.$$

The general form of the sine function is:

$$f(x) = A \sin \left[b \cdot (x + c) \right]$$

with three real parameters A, b, c. Usually one calls

A the amplitude
b the circular frequency
c the phase shift.

The amplitude describes the maximum deflection of the sine wave. In Fig. 2.12 the three sine functions $\sin x$, $2 \sin x$ and $\frac{1}{2} \sin x$ are compared. The change of the amplitudes from 1 to 2 or 1/2 has the consequence that the maximum deflection is doubled or halved.

The parameter b—the circular frequency —is a measure of the number of oscillations per interval. As Fig. 2.11 shows, the function $\sin x$ in the interval $[0, 2\pi]$ exactly one

2.2 Some Elementary Functions

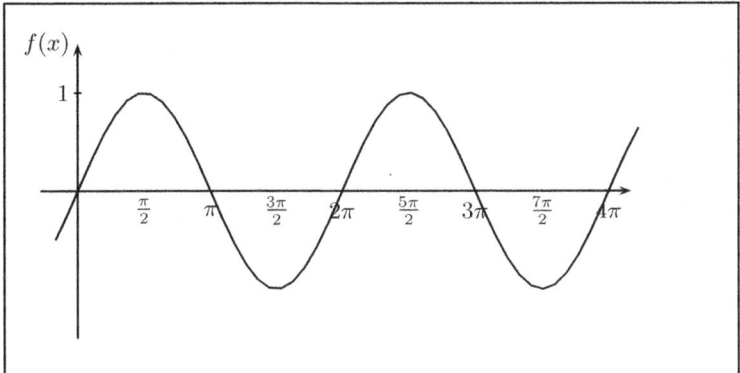

Fig. 2.11 The graph of the function $f(x) = \sin x$

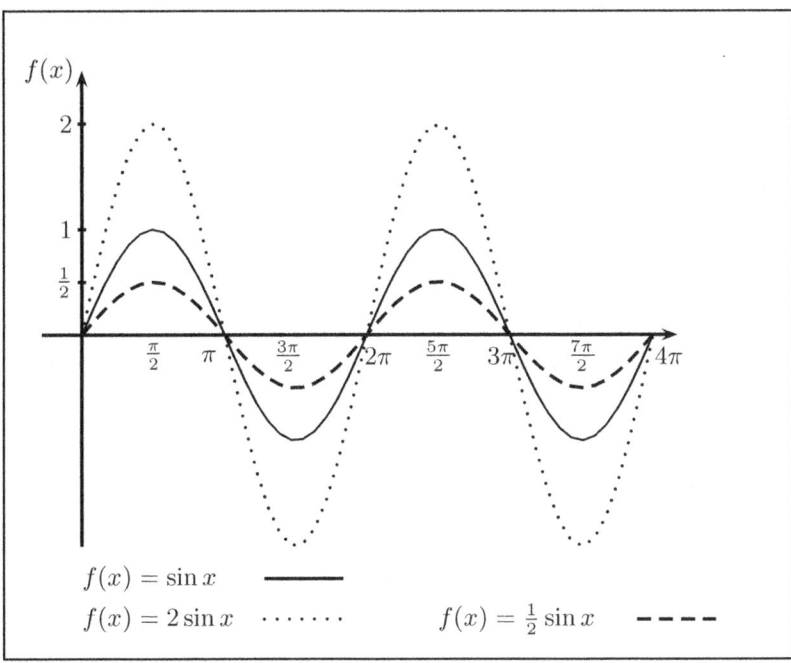

Fig. 2.12 The sine function with different amplitudes

complete oscillation. The function $\sin 2x$ performs two oscillations in this interval (Fig. 2.13). If one generalizes this for any b, then the function $\sin bx$ performs b oscillations in the interval $[0, 2\pi]$. The period length p of a sine wave results from the angular frequency with $p = 2\pi/b$. In Fig. 2.13 the three functions $\sin x$, $\sin 2x$ and $\sin x/2$ are shown.

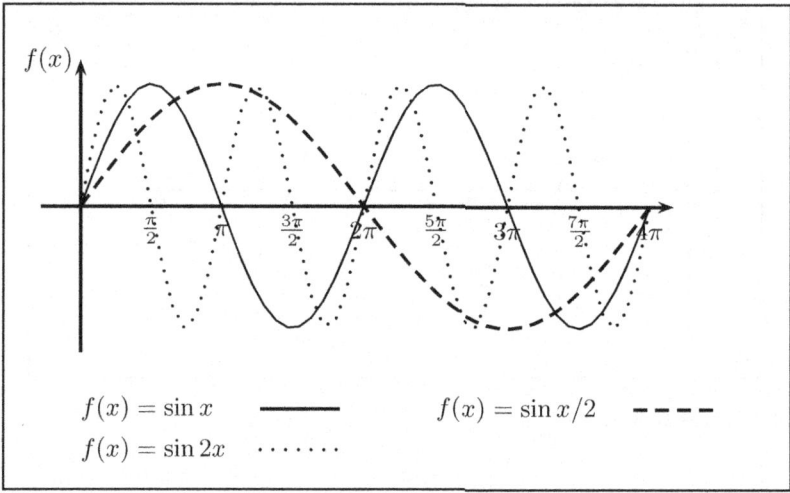

Fig. 2.13 The sine function with different frequencies

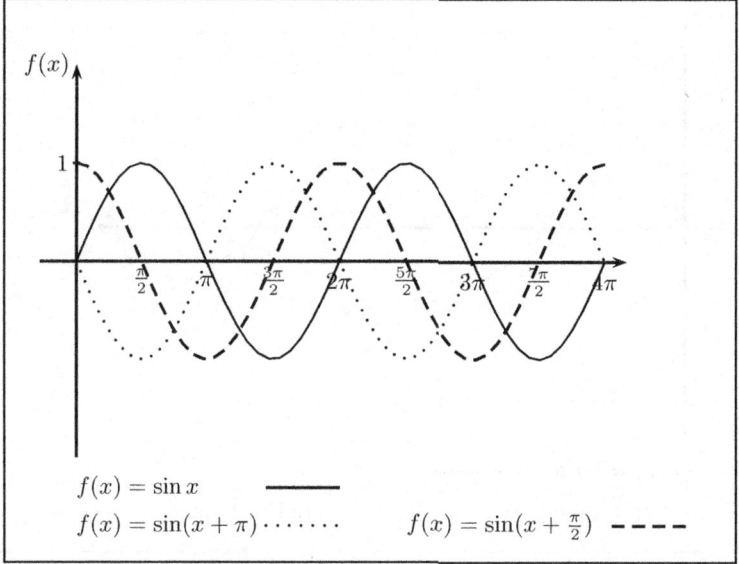

Fig. 2.14 The sine function with three different phases

The third parameter c describes a shift of the curve in the direction of increasing x for $c < 0$ and in the direction of decreasing x for $c > 0$. This phase shift is shown in Fig. 2.14 for displacements $\pi/2$ and π—this corresponds to a shift of 90 or 180 degrees.

2.2.11 Sectionally Defined Functions

If a single mapping rule is not applicable over the entire domain of definition, the domain of definition is divided into individual sections and a mapping rule is assigned to each section separately.

Examples The costs in production K depend on the amount produced x. This results in the course shown in Fig. 2.15.

The purchase of a machine increases the costs sharply. In addition, there are consumption costs for raw materials.

$$K(x) = \begin{cases} ax + b & \text{if } 0 \le x \le x_1 \\ ax + 2b & \text{if } x_1 < x \le x_2 \\ ax + 3b & \text{if } x_2 < x \le x_3 \end{cases}$$

In computers and digital data transmission, the character 'a' is represented by the bit sequence 01 100 001 in ASCII coding. So if you send the character a' over a data network, the bit pattern 01 100 001 is transmitted over the transmission medium. However, no 1s or 0s are transmitted, but a certain voltage (e.g. 5 volts) is applied to or not to the transmission channel. This bit pattern is transmitted in a certain time interval T.

Such a bit pattern can be mathematically formulated as follows:

$$h(t) = \begin{cases} 0 & \text{if } t \in [0, \frac{1}{8}T[\\ 1 & \text{if } t \in [\frac{1}{8}T, \frac{3}{8}T[\\ 0 & \text{if } t \in [\frac{3}{8}T, \frac{7}{8}T[\\ 1 & \text{if } t \in [\frac{7}{8}T, T] \end{cases}.$$

The sectionally defined function is shown in Fig. 2.16.

Fig. 2.15 Representation of the sectionally defined production costs

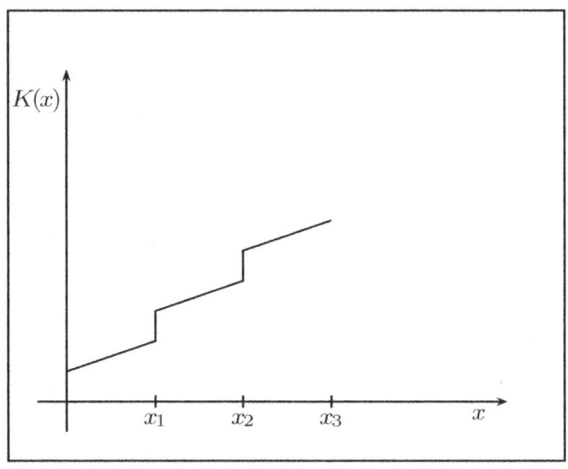

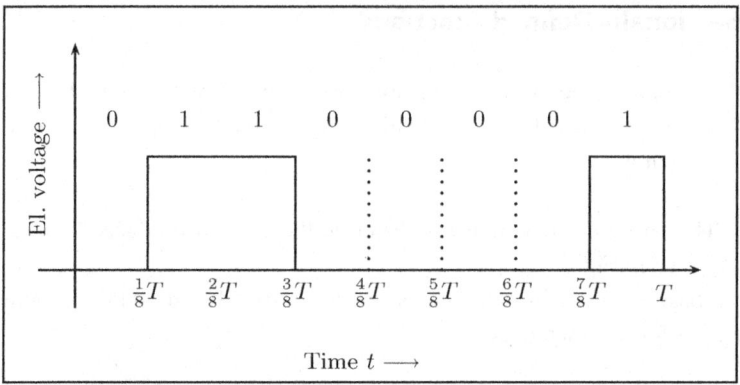

Fig. 2.16 The 01 100 001—bit pattern of the character 'a' in ASCII coding

2.2.12 Some Economic Functions

This section presents some functions used in economics by way of example. If one introduces an independent variable which is also called the **decision variable**, then the dependence of a size y on the decision variable x can often be represented as a functional relationship.

The formulation of such a relationship always has a model as its basis, in which a mapping of reality is carried out. A model simplifies reality by:

- the neglect of influences
- assumptions about the dependence of variables
- extrapolation of the range of validity.

How well a model describes reality must therefore always be validated retrospectively by plausibility considerations. In the following we restrict ourselves to functions with one variable. Functions with several variables are examined in Chap. 6.

- **Demand function**

  The **demand function** $x_N = x_N(p)$ describes the demanded quantity of a good in dependence on the price of the offered good or the corresponding inverse function, in which the price is expressed in dependence on the demanded quantity $p = p(x_N)$. The **Price-sales function** is also referred to as the Price-demand function. In a monopolistic market (only one supplier), it is assumed that the demand function is a monotonically decreasing function. This means that the higher the price of a good, the lower the demanded quantity. In Fig. 2.17 two monotonically decreasing curve progressions are shown.[4]

---

[4] In economic terms, such relationships are also represented by plotting price on the ordinate (y-axis) and the quantity demanded on the abscissa (x-axis).

2.2 Some Elementary Functions

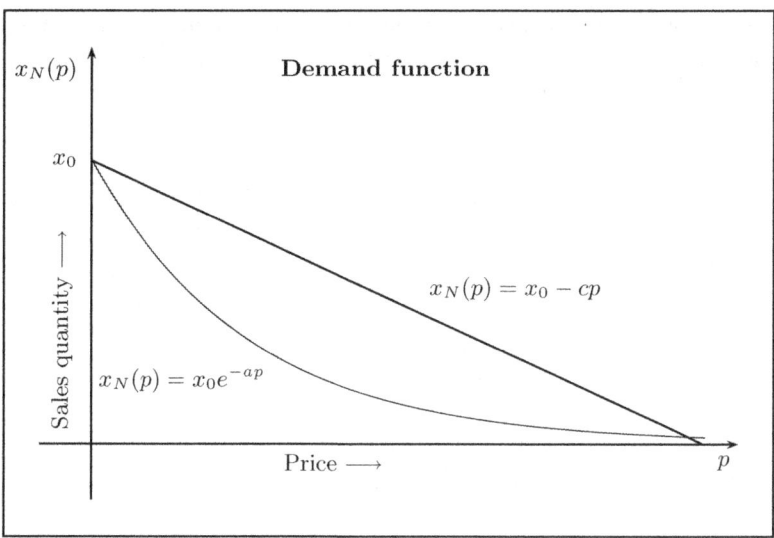

Fig. 2.17 Different forms of the demand function

1. A possible course is given by the linear function

$$x_N(p) = x_0 - c \cdot p \tag{2.4}$$

   x_0 and c are real constants. The parameter x_0 denotes the maximum demand at $p = 0$, if the good thus costs nothing. The point on the p-axis, through which the falling line runs, represents the price of the good, at which no demand is present anymore.

2. Another course is given by the function

$$x_N(p) = x_0 e^{-ap}, \quad a > 0$$

   with the real parameters x_0 and a. Here an exponential decrease of the demand with increasing price is assumed. At this point it should be pointed out that such a model is only meaningful for an upwardly limited price interval. An extrapolation for arbitrarily large prices is not possible.

3. There are also models in which the demand function is not monotonically decreasing. We speak of the *Snob effect* when a price increase leads to greater demand. For brand name products of higher quality, this may be the case.

In a homogeneous polypole (perfect market) the price is independent of the quantity, and thus a constant.

- **Supply function**
  The **supply function** gives the functional relationship between the market price p of a product and the quantity offered on the market x_A: $x = x_A(p)$ or $p = p(x_A)$. Since an

increasing market price increases the quantity offered by the producer, a monotonically increasing function is usually assumed here. Market equilibrium is achieved by setting the supply function equal to the demand function $x_N(p)$ (see Sect. 2.2.12) and thus determining the market price.

- **Revenue function**

  The revenue function describes the functional relationship between sales revenue and price. Since the relationship $E = x \cdot p$ exists between the price p, the quantity sold x and the corresponding revenue E, the *revenue* E can be represented as a function of the price p depending on the choice of the underlying price-sales function:

  $$E(p) = x(p) \cdot p.$$

  If a linear relationship is assumed in the price-sales function as in Eq. (2.4), it follows:

  $$\begin{aligned} E(p) &= x(p)p \\ &= (x_0 - cp)p \\ &= -cp^2 + x_0 p. \end{aligned} \qquad (2.5)$$

  Since it is a quadratic function that has a maximum at the vertex, there is an optimal price p_{opt} at which the revenue is maximal (cf. Fig. 2.18).

- **Production functions**

  Production functions describe the relationship between the:
  - Input r of a production (machine time, labor or another resource)
  - and the corresponding output (yield) x of the produced product.

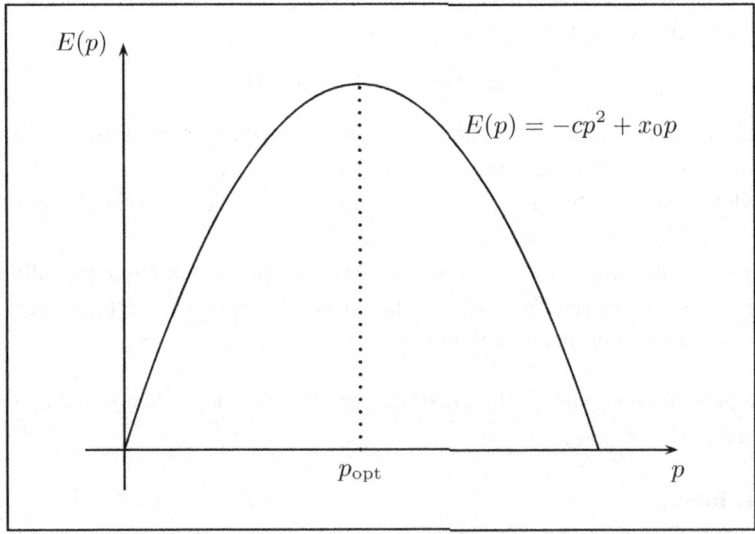

Fig. 2.18 The profit function $E(p)$ as a function of the price p

2.2 Some Elementary Functions

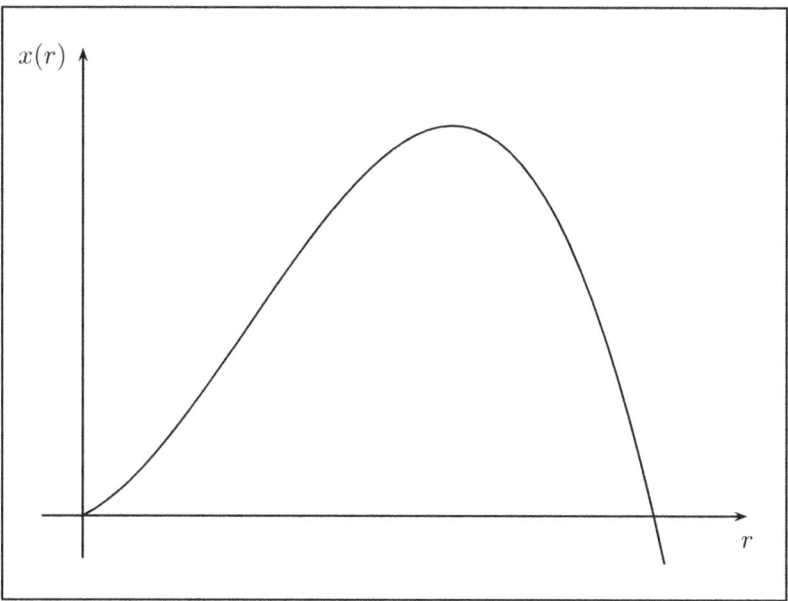

Fig. 2.19 The revenue-producing production function

Then
$$x = x(r); \quad r > 0$$
is the production function. Production functions are also referred to as yield function. We want to look at a series of models here that are realized by different production functions:

1. The yield-producing production function is based on a model in which, by using a resource, the revenue first increases disproportionately—starting at zero—then reaches a maximum and then decreases. A polynomial of 3rd degree can be used to represent such behaviour. We consider, for example:
$$x(r) = -r^3 + 7r^2 + 12r; \quad r > 0.$$

Note:
The application of seed to a specified area of land is a classic example of a revenue-producing production function (Fig. 2.19).

2. The **neoclassical production function** with the positive parameters a and c is given by
$$x(r) = c \cdot r^\alpha; \quad r > 0, 0 < \alpha < 1$$

The graph of this function with $a = 1/2$ is shown in Fig. 2.20.

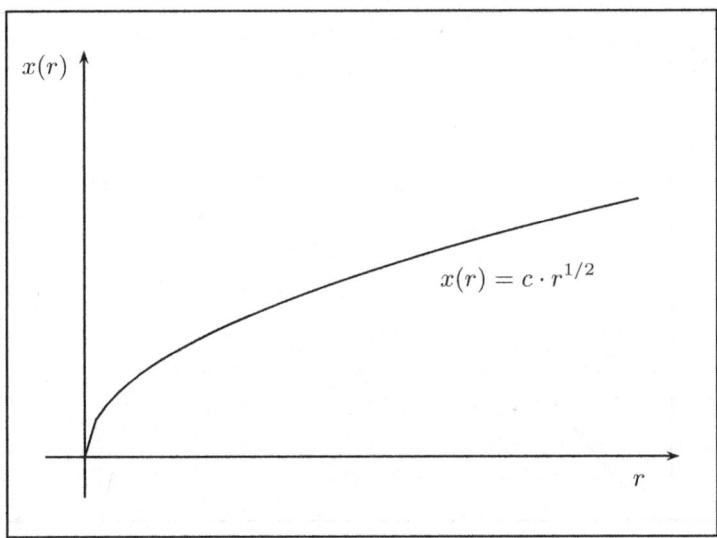

Fig. 2.20 The neoclassical production function

3. The **limitational production function** is defined by

$$x(r) = \begin{cases} c \cdot r & \text{if } r \leq r_0, \\ x_0 & \text{if } r > r_0 \end{cases}$$

with the two parameters c and x_0 (see Fig. 2.21). Here, there is initially a linear increase in output with the resource. From a certain point r_0 the output can no longer be increased by the use of the resource.

- **Cost function**
  The costs that arise in the production of an output x are recorded by the **cost function** $K(x)$. There is often a cost component KF(fixed costs) that does not depend on the amount of output x and one that depends on x (variable costs), $K_V(x)$. This gives the cost function the form

$$K(x) = K_F + K_V(x).$$

The average cost per unit of output x is called **unit cost**:

$$k(x) = \frac{K(x)}{x}.$$

2.3 The Inverse Function

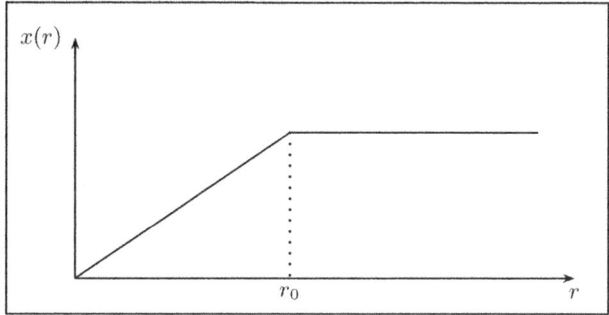

Fig. 2.21 The limitational production function

2.3 The Inverse Function

For a function:

$$f : D_f \longrightarrow W_f,$$
$$x \longmapsto y = f(x) \in W_f$$

for each x-value of the domain of definition D_f there corresponds *exactly* one y-value of the range of values W_f. The *uniqueness* is a characteristic property of a function. However, this does not exclude that the same y-value is assigned to different x-values, as can be seen in the diagram in Fig. 2.1.

For example, the quadratic function

$$f : \mathbb{R} \longrightarrow \mathbb{R},$$
$$x \longmapsto f(x) = x^2$$

assigns the same value $y \in \mathbb{R}$ to the two elements $x, -x \in \mathbb{R}$.

However, if the function

$$f : D_f \longrightarrow W_f,$$
$$x \longmapsto y = f(x) \in W_f$$

assigns different y values to different x values from the domain, then the inverse mapping $x = g(y)$ is also unique. It is created by solving the equation

$$y - f(x) = 0$$

for x (Fig. 2.22).

▶ **Definition (Inverse Function)** A function $y = f(x)$ with $x \in D_f$ and $y \in W_f$ is called invertible if there is exactly one $x \in D_f$ for each $y \in W_f$.

The mapping $g : y \to g(y)$ with $y \in W_f = D_g$ is called the **Inverse Function** of f. This function is often denoted by f^{-1}.

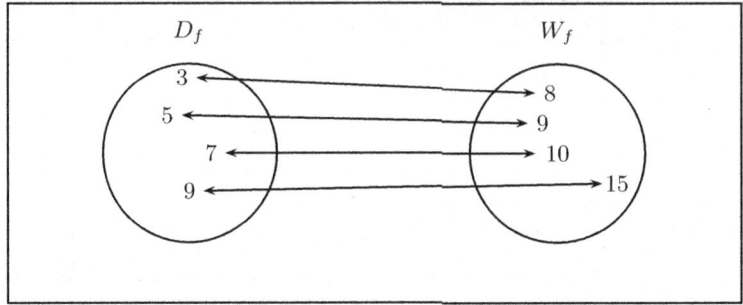

Fig. 2.22 Diagram of an invertible function. The function $y = f(x)$ assigns different y values from W_f to different x values from D_f

Remarks

1. For *every* function it holds:

$$x_1 = x_2 \implies f(x_1) = f(x_2); \text{ for all } x_1, x_2 \in D_f.$$

   For *reversible* functions it holds:

$$x_1 = x_2 \iff f(x_1) = f(x_2); \text{ for all } x_1, x_2 \in D_f.$$

2. If the graph of a function f is given, then the question of reversibility is easily decided. The function f is exactly reversible if every horizontal line intersects the function graph at most once.

Examples

1. Let

$$f : y = f(x) = 2x + 1, \quad x \in \mathbb{R}.$$

   To obtain the inverse function of this function, solving the equation

$$y = 2x + 1$$

   after x on

$$x = \frac{y}{2} - \frac{1}{2}.$$

   A renaming of the variables therefore leads to (Fig. 2.23)

$$f^{(-1)}(x) = \frac{x}{2} - \frac{1}{2}$$

2.3 The Inverse Function

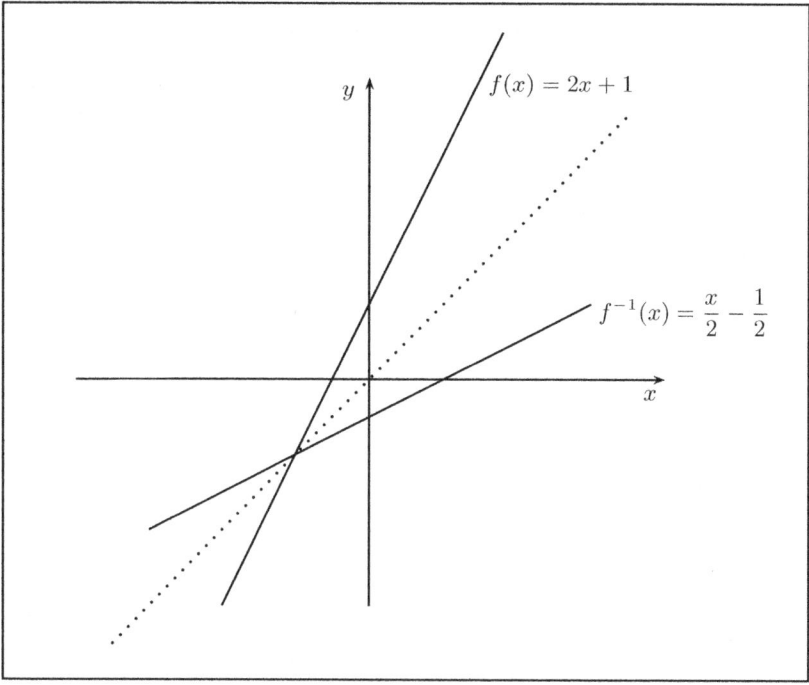

Fig. 2.23 The function $f(x) = 2x + 1$ and its inverse function

2. The function

$$f : \mathbb{R}^+ \longrightarrow \{y \mid 1 \leq y < \infty\},$$
$$x \longmapsto f(x) = x^2 + 1$$

is invertible with

$$f^{-1}(x) = +\sqrt{x - 1}.$$

3. The function

$$f : \{-\infty < x \leq 0\} \longrightarrow \{y \mid 1 \leq y < \infty\},$$
$$x \longmapsto f(x) = x^2 + 1$$

is invertible with

$$f^{-1}(x) = -\sqrt{x - 1}.$$

4. The function

$$f(x) = \sin x; \quad x \in [0, 4\pi]$$

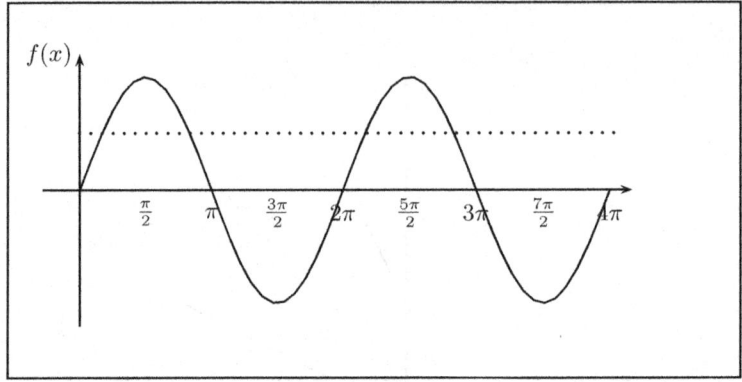

Fig. 2.24 The sine function with domain $D_f = [0, 4\pi]$

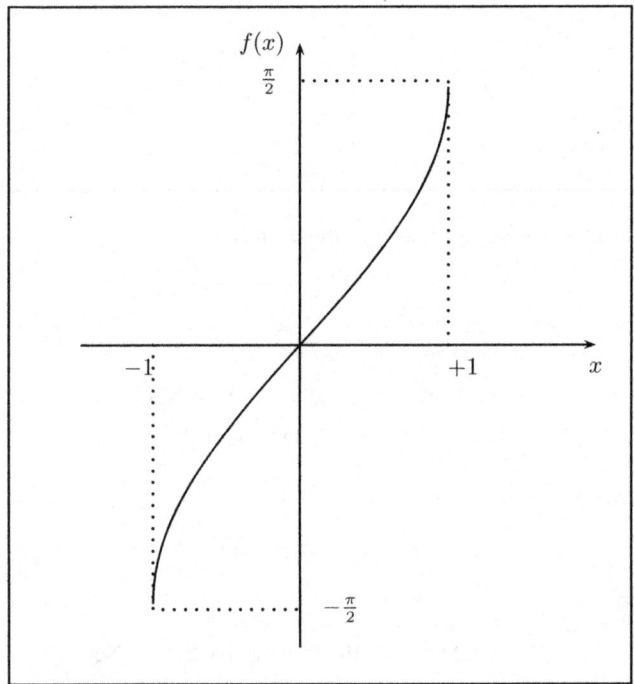

Fig. 2.25 The graph of the arc sine function $f(x) = \arcsin x$

is not reversible. As can be seen easily from Fig. 2.24, horizontal lines intersect the graph of this function multiple times. The domain is essential here. If one considers the sine function on the interval $[-\frac{\pi}{2}, +\frac{\pi}{2}]$, then a inverse function—that is the arcsin function—can be defined. The graph of this function is shown in Fig. 2.25, it is defined as follows:

$$f : \{-1 \leq x \leq +1\} \longrightarrow \left[-\frac{\pi}{2}, +\frac{\pi}{2}\right],$$
$$x \longmapsto f(x) = \arcsin x.$$

If function and inverse function are represented in the same coordinate system as in Fig. 2.23, the independent variable is always plotted on the abscissa. This in turn corresponds to a reflection at the 1st angle bisector.

2.4 Chained Functions

Functions of the form

$$g(x) = \sqrt{x^2 + 3x - 12}$$

or

$$k(x) = \exp\{-x^2\}$$

can be thought of as resulting from the *consecutive execution* of two elementary functions. The resulting function terms arise from substituting one function term into the other.

The above functions arise from the consecutive execution of:

$$f(x) = x^2 + 3x - 12$$
$$h(f) = \sqrt{f}$$
$$h(f(x)) = \sqrt{f(x)} = \sqrt{x^2 + 3x - 12} = g(x)$$

or

$$f(x) = -x^2$$
$$h(f) = \exp\{f\}$$
$$h(f(x)) = \exp\{f(x)\} = \exp\{-x^2\} = k(x).$$

The consecutive execution of functions is called **chaining**.

Example If

$$f(x) = 2x + 2 \quad \text{and} \quad g(y) = y^2$$

then the chaining of these two functions is:

$$g(f(x)) = (f(x))^2 = (2x + 2)^2 = 4x^2 + 4x + 4.$$

Remarks

1. In general, the chaining of functions is not commutative, which means:
$$g(f(x)) \neq f(g(x)).$$

2. In order for a chaining $g(f(x))$ of two functions
$$f : D_f \longrightarrow W_f$$
and
$$g : D_g \longrightarrow W_g$$
to be possible, the range of the function f must not be disjoint from the domain of the function g:
$$W_f \cap D_g \neq \emptyset.$$

3. More than two functions can also be chained.
4. If f^{-1} is the inverse function of f, then:
$$f(f^{-1}(x)) = x \quad \text{and} \quad f^{-1}(f(x)) = x.$$

This means that the chaining of a function with its inverse function is commutative and the chaining represents the identical mapping:
$$x \longmapsto x.$$

2.5 Properties of Functions

We consider some basic properties of real functions in this section, such as:

- Limitation
- Monotonicity
- Symmetry
- Injectivity, surjectivity and bijectivity.

2.5.1 Limitation

▶ **Definition (Limitation of Functions)** A function f with domain D_f is called limited from above, if there is a number $k \in \mathbb{R}$ with
$$f(x) \leq k \quad \text{for all } x \in D_f.$$

The number $k \in \mathbb{R}$ is called the **upper limit**. Analogously, f is called bounded from below, if there is a number $k' \in \mathbb{R}$ with

2.5 Properties of Functions

$$f(x) \geq k' \quad \text{for all } x \in D_f.$$

The number k' is called the **lower limit**. A function

$$f : D_f \longrightarrow W_f$$

is called limited, if f is limited both from above and from below.

Examples The function

$$f(x) = -\frac{1}{x^2} + 6$$

is limited from above, because f never exceeds the value $k = 6$, i.e. $f(x) \leq 6$ for all $x \in D_f$.

The function

$$f(x) = A \exp\{-x^2\}$$

is limited, because $f(x) \leq A$ and $f(x) \geq 0$ for all $x \in D_f = \mathbb{R}$.

2.5.2 Monotonicity

An important property of functions is when the function values always increase or decrease with increasing argument values. We have already seen examples of such functions, for example, the logarithm function considered in Sect. 2.2.9 grows to ever larger values with increasing argument. Such functions are called **strictly monotonic increasing** or **strictly monotonic decreasing**.

▶ **Definition (Monotonicity of Functions)** Analogously a function f is called **strictly monotonically increasing** in an interval $I \subset D_f$, if for all $x_1, x_2 \in I$ with $x_1 < x_2$ it holds: $f(x_1) < f(x_2)$.

If one allows the equality of function values, then the attribute 'strictly' is omitted.

Example The function

$$f : \mathbb{R} \longrightarrow \mathbb{R},$$
$$x \longmapsto f(x) = x^3$$

is strictly monotonic on $D_f = \mathbb{R}$, because if $x_1 < x_2$, then $f(x_1) < f(x_2)$.

For strictly monotonic functions it can be shown that an inverse function always exists.

2.5.3 Symmetry

We consider in this section two easily recognizable symmetries of functions, the axis symmetry to the y-axis and the point symmetry to the origin.

A function f with the property
$$f(-x) = f(x)$$
is called **even**. The function graph is mirror-symmetrical to the y-axis.

Example The function
$$f : \mathbb{R} \longrightarrow \mathbb{R},$$
$$x \longmapsto f(x) = x^2$$
satisfies $f(-x) = f(x)$ and is therefore an even function.

A function f with the property
$$f(-x) = -f(x)$$
is called **odd**. The function graph is point-symmetric to the origin of the coordinate system.

Example The function
$$f : \mathbb{R} \longrightarrow \mathbb{R},$$
$$x \longmapsto f(x) = x^3$$
satisfies $f(-x) = -f(x)$ and is therefore an odd function.

An example of a symmetric function is given by
$$f : \mathbb{R} \longrightarrow \mathbb{R}$$
$$x \longrightarrow f(x) = x^4 - 2x^2 + 3$$
with $f(-x) = f(x)$. The function graph is shown in Fig. 2.26. It holds
$$\begin{aligned} f(-x) &= (-x)^4 - 2(-x)^2 + 3 \\ &= (-1)^4 x^4 - 2(-1)^2 x^2 + 3 \\ &= x^4 - 2x^2 + 3 \\ &= f(x). \end{aligned}$$

Coordinate Transformations

Some functions can be transformed into an elementary function by a simple coordinate transformation, which means a shift of the coordinate system. We have already made use of this in the phase shift of the sine function. At this point we want to look at this in general.

2.5 Properties of Functions

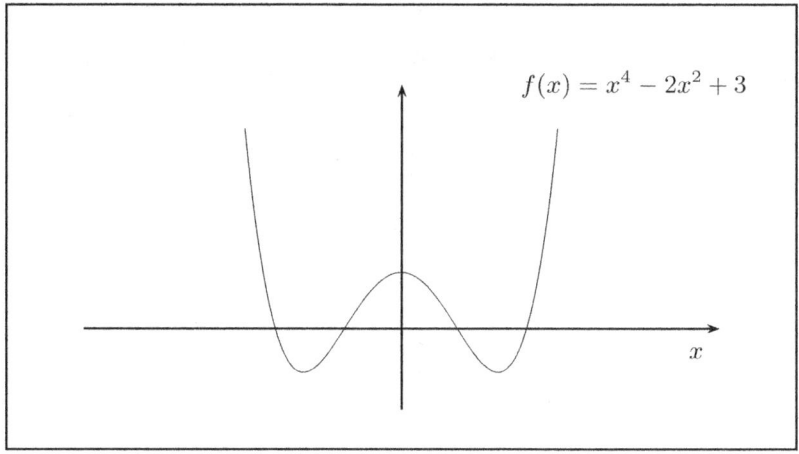

Fig. 2.26 Graph of the function: $f(x) = x^4 - 2x^2 + 3$

Given a function $y = f(x)$, which can be brought into the form

$$y = f(x - x_0) + y_0.$$

With the coordinate transformation

$$x - x_0 = \hat{x}$$
$$y - y_0 = \hat{y}$$

one obtains $\hat{y} = f(\hat{x})$.

In the new coordinate system, the easily recognizable symmetries to the coordinate axis and to the origin can be investigated again. As an example, we consider the function

$$y = \frac{1}{x-1} + 2$$
$$\Longleftrightarrow \quad y - 2 = \frac{1}{x-1}.$$

With $x - 1 = \hat{x}$ and $y - 2 = \hat{y}$, the hyperbolic function $\hat{y} = \frac{1}{\hat{x}}$ results. Figure 2.27 shows the graph of the hyperbolic function in the two coordinate systems.

2.5.4 Injectivity, Surjectivity and Bijectivity

Definition
Let M_1, M_2 be two sets and let

$$f : M_1 \longrightarrow M_2$$

be a function.

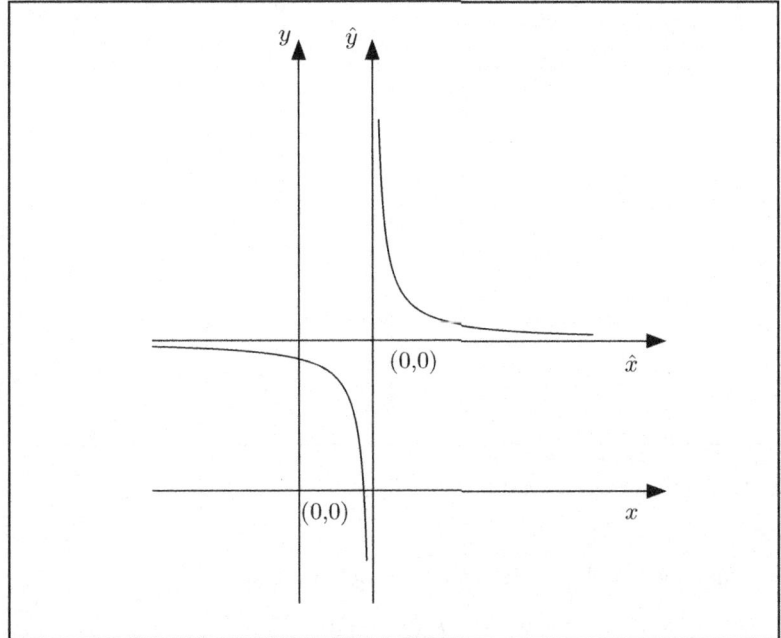

Fig. 2.27 Shift of the coordinate system

1. The function f is called **injective** or *reversible uniquely* or also *uniquely*, if from $x_1 \neq x_2$ follows $f(x_1) \neq f(x_2)$ for all $x_1, x_2 \in M_1$.
2. The function f is called **surjective**, if for each $y \in M_2$ an $x \in M_1$ with $y = f(x)$ exists.
3. The function f is called **bijective** if and only if f is injective and surjective.

Remarks

(a) If a function f is injective and $x_1 \neq x_2$, then $f(x_1) \neq f(x_2)$, i.e. different images have different primal images. To show that a function is injective, it is often easier to show the following equivalent statement:

If $f(x_1) = f(x_2)$, then $x_1 = x_2$.

To show that a function *is not* injective, it is enough to give a counter-example.
(b) If a function f is surjective, one also says that f is a mapping from M_1 **to** the set M_2. With a surjective function, the potential image range is exhausted.
(c) If the function $f: M_1 \to M_2$ is bijective, then for every $y \in M_2$ there is exactly one $x \in M_1$ with $y = f(x)$.

2.6 Limits

Before we look at the limit of a function, we first examine this concept for sequences and series.

2.6.1 Convergence and Limits of Sequences and Series

▶ **Definition (Cauchy Convergence Criterion)** Let $(a_n)_{n \in \mathbb{N}}$ be a sequence (see Sect. 1.3). We say that the sequence $(a_n)_{n \in \mathbb{N}}$ converges to the limit (or limit) $a \in \mathbb{R}$ if the following is true: For any arbitrarily small $\varepsilon > 0$ there is a sequence index n_0 such that

$$|a_n - a| < \epsilon \quad \text{for all } n > n_0. \tag{2.6}$$

For this we write (lim stands for *limit*)

$$\lim_{n \to \infty} a_n = a.$$

Intuitively, the convergence of a sequence $(a_n n \geq 1)$ means that the sequence elements a_n lie in a stripe of width 2ε around the limit value a for a value n_0; the value ε can be made arbitrarily small. This is illustrated in Fig. 2.28.

We call a sequence that converges to the value 0 a **null sequence**. A sequence that does not converge is called a divergent sequence.

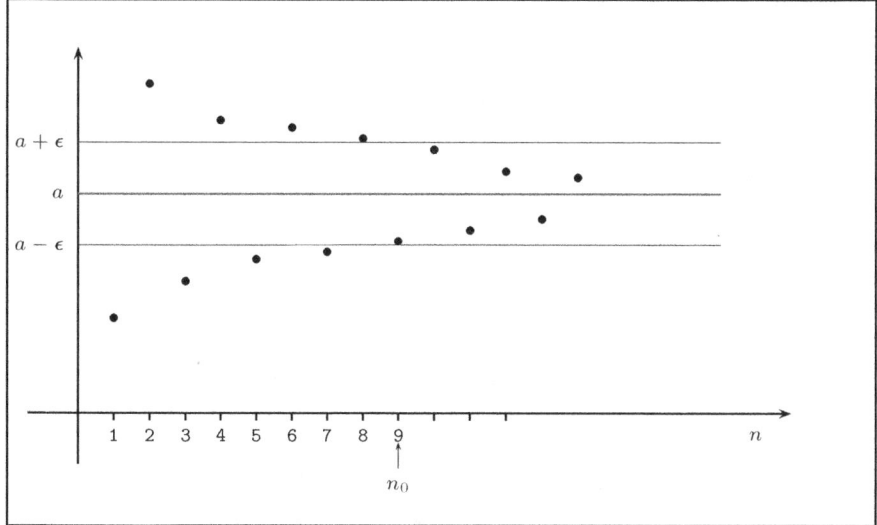

Fig. 2.28 Convergence of a sequence

Examples

1. The sequence

$$a_n = 1 + \frac{1}{n^2}$$

converges to the value $a = 1$ for $n \to \infty$, so

$$\lim_{n\to\infty} a_n = \lim_{n\to\infty} \left(1 + \frac{1}{n^2}\right) = 1.$$

Because: Let an arbitrary $\varepsilon > 0$ be given. Then

$$\left|1 + \frac{1}{n^2} - 1\right| = \left|\frac{1}{n^2}\right| < \epsilon.$$

So:

$$\frac{1}{n^2} < \epsilon \implies n > n_0 = \frac{1}{\sqrt{\epsilon}}, n \in \mathbb{N}.$$

Here n_0 is the smallest integer greater than or equal to $1/\sqrt{\epsilon}$. This shows that if an arbitrary $\varepsilon > 0$ is given, then for all

$$n > n_0 = \frac{1}{\sqrt{\epsilon}}$$

the distance from a_n and $a = 1$ is less than this ε:

$$|a_n - 1| < \epsilon.$$

For example, if $\epsilon = \frac{1}{400}$, then for all indices $n > 20 = n_0$ it holds that

$$\frac{1}{n^2} < \frac{1}{400}.$$

The sequence $\frac{1}{n^2}$ is a null sequence.

2. The harmonic sequence

$$1, \frac{1}{2}, \frac{1}{3}, \ldots, \frac{1}{n}, \ldots \qquad (2.7)$$

is a null sequence, i.e. the sequence converges to the value 0:

$$\lim_{n\to\infty} \frac{1}{n} = 0.$$

3. The sequence

$$a_n = \sqrt[n]{n}, \qquad (2.8a)$$

converges to 1.

2.6 Limits

Proof with Cauchy Criterion

For any arbitrary ε with $0 < \varepsilon < 1$ we construct a value $n0 \in N$ in way that for each $n > n0$ the inequality is true:
Then

$$|\sqrt[n]{n} - 1| < \epsilon. \tag{2.8b}$$

$$\sqrt[n]{n} < 1 + \epsilon.$$

Raising the left and right side of the equation to a power and applying the binomial theorem yields:

$$n < (1+\epsilon)^n = \sum_{k=0}^{n} \binom{n}{k} \epsilon^k 1^{n-k} = \sum_{k=0}^{n} \binom{n}{k} \epsilon^k. \tag{2.8c}$$

We first show that from

$$n < 1 + \frac{n}{2}(n-1)\epsilon^2 \tag{2.8d}$$

the inequality (2.8c) follows: If we add only positive terms to the right side of the inequality, we get:

$$n < 1 + \frac{n}{2}(n-1)\epsilon^2 \implies n < 1 + n\epsilon + \frac{n}{2}(n-1)\epsilon^2$$
$$+ \frac{n}{6}(n-1)(n-2)\epsilon^3 + \cdots$$
$$\iff n < \sum_{k=0}^{n} \binom{n}{k}\epsilon^k$$
$$\iff n < (1+\epsilon)^n.$$

From the inequality (2.8d) we get:

$$n < 1 + \frac{n}{2}(n-1)\epsilon^2 \iff n - 1 < \frac{n}{2}(n-1)\epsilon^2$$
$$\iff \frac{2}{n} < \epsilon^2$$
$$\iff n > \frac{2}{\epsilon^2}.$$

Thus we have shown: For each $\varepsilon > 0$ there exists a n_0, so that for all $n > n_0$ with

$$n > \frac{2}{\epsilon^2}$$

the inequality (2.8b) is fulfilled. According to the Cauchy convergence criterion, the sequence (2.8a) therefore converges to 1.

Calculating with Limits

With the help of the definition of convergence Eq. (2.6) it can be shown that limits and basic arithmetic are compatible. This means that if, for example, you have two sequences $(a_n)_{n \geq 1}$ and $(b_n)_{n \geq 1}$, which converge to a and b respectively, then the question is what is the limit of the sequence $a_n \cdot b_n$. The answer to this is given by the following calculation rules, which are also called limit theorems. The limit theorems provide the tools to calculate limits of more complicated sequences. This also has applications in the limits of functions.

The following calculation rules apply to limits:[5]

Calculation Rules for Limits, Limit Theorems

Let $(a_n)_{n \geq 1}$ and $(b_n)_{n \geq 1}$ be convergent sequences in $\mathbb{R}$ with the limits

$$\lim_{n \to \infty} a_n = a, \quad \lim_{n \to \infty} b_n = b.$$

Then holds:

1. Linearity:

$$\lim_{n \to \infty} (a_n \pm b_n) = \lim_{n \to \infty} a_n \pm \lim_{n \to \infty} b_n = a \pm b. \tag{2.9a}$$

2. For all $c \in \mathbb{R}$ it holds:

$$\lim_{n \to \infty} (c \cdot a_n) = c \cdot \lim_{n \to \infty} a_n = c \cdot a. \tag{2.9b}$$

3. Compatibility with multiplication:

$$\lim_{n \to \infty} (a_n \cdot b_n) = \left(\lim_{n \to \infty} a_n\right) \cdot \left(\lim_{n \to \infty} b_n\right) = a \cdot b. \tag{2.9c}$$

4. Compatibility with division:
   If the limit $b \neq 0$ is, then

$$\lim_{n \to \infty} \left(\frac{a_n}{b_n}\right) = \frac{\lim_{n \to \infty} a_n}{\lim_{n \to \infty} b_n} = \frac{a}{b}. \tag{2.9d}$$

5. For *integers* $p, q > 0$ it holds:

$$\lim_{n \to \infty} a_n^{p/q} = \left(\lim_{n \to \infty} a_n\right)^{p/q} = a^{\frac{p}{q}}. \tag{2.9e}$$

Example We investigate the sequence

$$a_n = \frac{4n^2 - 2n + 2}{n^2 + 2n - 2} \tag{2.10}$$

---

[5] To prove, see Arens et al. (2018), Sect. 6.3.

2.6 Limits

on its behavior for $n \to \infty$ with the help of the limit theorems. So we are looking for

$$\lim_{n \to \infty} \left(\frac{4n^2 - 2n + 2}{n^2 + 2n - 2} \right).$$

Note that the rule (2.9d) **cannot** be applied here, because the limit of the numerator and the limit of the denominator do not exist.[6] This is not consistent with the assumptions of the limit theorems. We cancel the fraction (2.10) with the highest power that occurs in the numerator and denominator; this results in terms with zero sequences, which can be processed within the framework of the limit theorems.

$$\begin{aligned} a_n &= \frac{4n^2 - 2n + 2}{n^2 + 2n - 2} \\ &= \frac{n^2 \left(4 - \frac{2}{n} + \frac{2}{n^2}\right)}{n^2 \left(1 + \frac{2}{n} - \frac{2}{n^2}\right)} \\ &= \frac{4 - \frac{2}{n} + \frac{2}{n^2}}{1 + \frac{2}{n} - \frac{2}{n^2}}. \end{aligned}$$

The numerator and denominator now each consist of a constant term and zero sequences. Therefore, the limit theorems can be applied, in particular we obtain with the rule (2.9d):

$$\begin{aligned} \lim_{n \to \infty} a_n &= \lim_{n \to \infty} \frac{4 - \frac{2}{n} + \frac{2}{n^2}}{1 + \frac{2}{n} - \frac{2}{n^2}} \\ &= \frac{\lim_{n \to \infty} \left(4 - \frac{2}{n} + \frac{2}{n^2}\right)}{\lim_{n \to \infty} \left(1 + \frac{2}{n} - \frac{2}{n^2}\right)}. \end{aligned}$$

If the rule (2.9a) is applied, the following is obtained:

$$\lim_{n \to \infty} a_n = \frac{\lim_{n \to \infty} 4 - \lim_{n \to \infty} \frac{2}{n} + \lim_{n \to \infty} \frac{2}{n^2}}{\lim_{n \to \infty} 1 + \lim_{n \to \infty} \frac{2}{n} - \lim_{n \to \infty} \frac{2}{n^2}}.$$

It follows that the numerator converges to 4, the denominator converges to 1:

$$\lim_{n \to \infty} a_n = \frac{4}{1} = 4.$$

---

[6] We say for the case of a divergent sequence that its limit does not exist. This is the case in the current example with the sequence of numbers $4n^2 - 2n + 2$ in the numerator and the sequence of numbers $n^2 + 2n - 2$ in the denominator.

Convergence of a Series

In Sect. 1.3 we introduced partial sums s_n. The finite sum of terms of a sequence is a **finite series**:

$$s_1 = a_1,$$
$$s_2 = a_1 + a_2,$$
$$s_3 = a_1 + a_2 + a_3,$$
$$\vdots$$
$$s_n = a_1 + a_2 + \cdots + a_n = \sum_{i=1}^{n} a_i$$

At first it does not seem clear what it means to consider an infinite sum

$$a_1 + a_2 + a_3 + \cdots$$

since we do not know how to add an infinite number of numbers. However, if the sequence of partial sums $(s_n)_{n \geq 1}$ goes to a limit for $n \to \infty$, then we say that the (infinite) sum of the series converges. We define that this infinite sum is the limit. If it is not the case that the series converges, then we call the series divergent. If the series converges, then the value of the series is:

$$\sum_{i=1}^{\infty} a_i = \lim_{n \to \infty} s_n = \lim_{n \to \infty} (a_1 + a_2 + \cdots + a_n).$$

Convergence Behavior of the Geometric Series

Consider the geometric sequence

$$1, \frac{1}{2}, \frac{1}{4}, \frac{1}{8}, \frac{1}{16}, \ldots$$

We form the partial sums

$$s_n = 1 + \frac{1}{2} + \frac{1}{4} + \cdots + \frac{1}{2^n}.$$

These partial sums tend to a limit, which has the value 2. This can be seen as follows. We set $q = \frac{1}{2}$, then it is:

$$s_n = 1 + q + q^2 + \cdots + q^n,$$
$$q \cdot s_n = q + q^2 + q^3 + \cdots + q^{n+1}.$$

If one forms the difference of these two finite sequences, then exactly two terms remain:

$$s_n - q s_n = 1 - q^{n+1}.$$

2.6 Limits

It follows that:

$$S_n = \frac{1-q^{n+1}}{1-q} = \frac{1}{1-q} - \frac{q^{n+1}}{1-q}. \qquad (2.11)$$

This gives us the sum value of a finite geometric series (r is any factor not equal to 0) to

$$\boxed{r + qr + rq^2 + \cdots + rq^n = r \cdot \sum_{i=1}^{n} q^i = r\frac{1-q^{n+1}}{1-q}.} \qquad (2.12)$$

If n becomes very large, the term q^{n+1} in Eq. (2.11) tends to 0. This gives

$$\lim_{n\to\infty} s_n = \lim_{n\to\infty} \sum_{i=0}^{n} q^i = \sum_{i=0}^{\infty} q^i = \frac{1}{1-q}.$$

For $q = 1/2$ the limit is 2. Note that this argument is not only valid for $q = 1/2$, because for any value $-1 < q < 1$ the term q^{n+1} tends to 0 when $n \to \infty$.

Convergence behavior of the harmonic series[7]

We consider the harmonic series

$$1 + \frac{1}{2} + \frac{1}{3} + \cdots + \frac{1}{n}$$

The terms of this series become smaller and smaller—they form a null sequence—but the harmonic series is divergent. It grows very slowly over all limits. This implies that the requirement that the terms of a series become smaller and smaller in order for a series to converge is a necessary but not a sufficient condition.

The proof[8] goes back to Nicolae Oresme (1323–1382). We write the harmonic series in the form

$$S = 1 + \frac{1}{2} + \underbrace{\frac{1}{3} + \frac{1}{4}}_{} + \underbrace{\frac{1}{5} + \frac{1}{6} + \frac{1}{7} + \frac{1}{8}}_{} + \cdots \qquad (2.13a)$$

In each of the bracketed groups of terms we replace the first terms by the respective last term. This procedure generates a new sequence

$$S' = 1 + \frac{1}{2} + \underbrace{\frac{1}{4} + \frac{1}{4}}_{} + \underbrace{\frac{1}{8} + \frac{1}{8} + \frac{1}{8} + \frac{1}{8}}_{} + \cdots \qquad (2.13b)$$

---

[7] See the book by Maor (2017).
[8] See Maor (2017), Appendix or Merzbach and Boyer, p. 242.

In this replacement we have replaced each term of the original series by an equal or smaller term. It follows that each partial sum of S'—that is S'_n—is smaller than the corresponding partial sum of S which we denote by S_n.

$$S_n > S'_n. \tag{2.13c}$$

We can write the series S' in Eq. (2.13b) as

$$S' = 1 + \frac{1}{2} + \frac{1}{2} + \frac{1}{2} + \frac{1}{2} + \frac{1}{2} + \cdots, \tag{2.13d}$$

since each bracketed group of numbers in the series (2.13b) yields the value 1/2. Obviously, the series (2.13d) diverges, due to the condition (2.13c), that each partial sum of the harmonic series is greater than S'_n, the harmonic series must also diverge.

Convergence criteria for series

The question arises as to how one can decide whether a series converges. Here we consider two convergence criteria.[9] Consider the infinite series

$$S = \sum_{i=1}^{\infty} a_n. \tag{2.14}$$

1. **Quotient Criterion**

   The infinite series (2.14) converges (absolutely), if

   $$\lim_{n \to \infty} \left| \frac{a_{n+1}}{a_n} \right| < 1. \tag{2.15}$$

   The series (2.14) diverges, if

   $$\lim_{n \to \infty} \left| \frac{a_{n+1}}{a_n} \right| > 1.$$

   The case

   $$\lim_{n \to \infty} \left| \frac{a_{n+1}}{a_n} \right| = 1$$

is not decided.

---

[9] There are other convergence criteria in addition to the methods presented here. It should be noted that there is no royal road that decides which criterion to apply to which type of series in order to show convergence. For further discussion of this topic, see Arens et al. (2018), Chap. 8, in particular the overview in Sect. 8.4.

2.6 Limits

Example Consider the infinite series

$$S = \sum_{n=1}^{\infty} \frac{n!}{n^n}. \tag{2.16}$$

Now

$$\frac{a_{n+1}}{a_n} = \frac{(n+1)! n^n}{(n+1)^{n+1} n!} = \frac{(n+1) n^n}{(n+1)^{n+1}} = \left(\frac{n}{n+1}\right)^n$$

$$= \frac{1}{(1+\frac{1}{n})^n} \leq \frac{1}{2} < 1.$$

Thus the series (2.16) converges. This situation justifies the statement that n^n grows stronger than $n!$

2. **Root criterion**

The series (2.14) converges (absolutely) for

$$\lim_{n \to \infty} \sqrt[n]{|a_n|} < 1. \tag{2.17}$$

If the expression on the left side >1, the series diverges. If the limit is equal to 1, no statement can be made.

Example Let

$$S = \sum_{i=1}^{\infty} \frac{1}{n^n} \quad \text{where } a_n = \frac{1}{n^n}.$$

Then

$$\lim_{n \to \infty} \sqrt[n]{a_n} = \lim_{n \to \infty} \sqrt[n]{\frac{1}{n^n}} = \lim_{n \to \infty} \frac{1}{n} = 0 < 1. \tag{2.18}$$

Therefore, the series (2.18) converges.

Sequences and series can also be formed by function terms. A series of the form

$$\sum_{i=1}^{\infty} f_i(x)$$

is called Function series. Then

$$S_n = \sum_{i=1}^{n} f_i(x)$$

is the partial sum of the function series. The most important function series are the **power series**. These have the form

$$P(x) = \sum_{n=0}^{\infty} a_n x^n \quad \text{or} \quad P(x) = \sum_{n=0}^{\infty} a_n (x - x_0)^n, \tag{2.19}$$

with $a_n \in \mathbb{R}$ for all n.

The convergence of a power series is dependent on the values that the argument x can assume. The set of all $x \in \mathbb{R}$, for which the power series $P(x)$ converges, is called the **convergence radius** r of the power series $P(x)$. It holds that:

$P(x)$ converges for $|x| < r$.
$P(x)$ diverges for $|x| > r$.

For $|x| = r$ no general statements about the convergence of a power series can be made. If a power series converges only for one point, e. g. $x = 0$, then $r = 0$. With the convergence criteria for series one obtains the convergence radius r. From the quotient criterion (2.15) one obtains:

$$r = \lim_{n \to \infty} \left| \frac{a_n}{a_{n+1}} \right|. \tag{2.20}$$

From the root criterion (2.17) one obtains:

$$r = \lim_{n \to \infty} \left(\sqrt[n]{|a_n|} \right)^{-1}. \tag{2.21}$$

Examples Consider the power series

$$P(x) = \sum_{n=0}^{\infty} \frac{x^n}{n!}.$$

Then according to the criterion (2.20):

$$r = \lim_{n \to \infty} \left| \frac{a_n}{a_{n+1}} \right| = \lim_{n \to \infty} \frac{(n+1)!}{n!} = \lim_{n \to \infty} (n+1) = \infty.$$

This means that the radius of convergence is arbitrarily large, the power series (Sect. 2.6.1) converges for all $x \in \mathbb{R}$.

Consider the power series

$$P(x) = \sum_{n=0}^{\infty} \frac{n}{2^n} x^n. \tag{2.22}$$

2.6 Limits

If you apply the criterion (2.21), you get:

$$r = \lim_{n \to \infty} \frac{1}{\sqrt[n]{|a_n|}}$$

$$= \lim_{n \to \infty} \frac{1}{\sqrt[n]{\frac{n}{2^n}}}$$

$$= \lim_{n \to \infty} \frac{2}{\sqrt[n]{n}}$$

$$= 2 \cdot \lim_{n \to \infty} \frac{1}{\sqrt[n]{n}} = 2.$$

The series (2.22) converges on the open interval $]-2, +2[$.[10]

2.6.2 The Limit Concept for Functions

For some functions, the behavior of the function values when the independent variable approaches a certain value is of interest. This is particularly relevant if the function is sectionally defined or undefined at certain points.

Under a **limit of a function** one understands the behavior of a function $f(x)$, when the variable x approaches a certain value x_0 of the domain of definition. One also writes $x \to x_0$ for this.

The function value to which the function approaches for $x \to x_0$ is called the limit g and one writes for this:

$$\lim_{x \to x_0} f(x) = g.$$

Read: 'Limes $f(x)$ for x approaching x_0'.
In many cases, this consideration does not lead to new knowledge.
For example, if

$$f(x) = 2x^2 + 1,$$

then

$$\lim_{x \to 1} f(x) = 2 \cdot 1 + 1 = 3.$$

Here one can insert the value $x = 1$ directly.

---

[10] Note that with Eq. (2.8a) we have shown that the limit $n \to \infty$ of $\sqrt[n]{n}$ is 1.

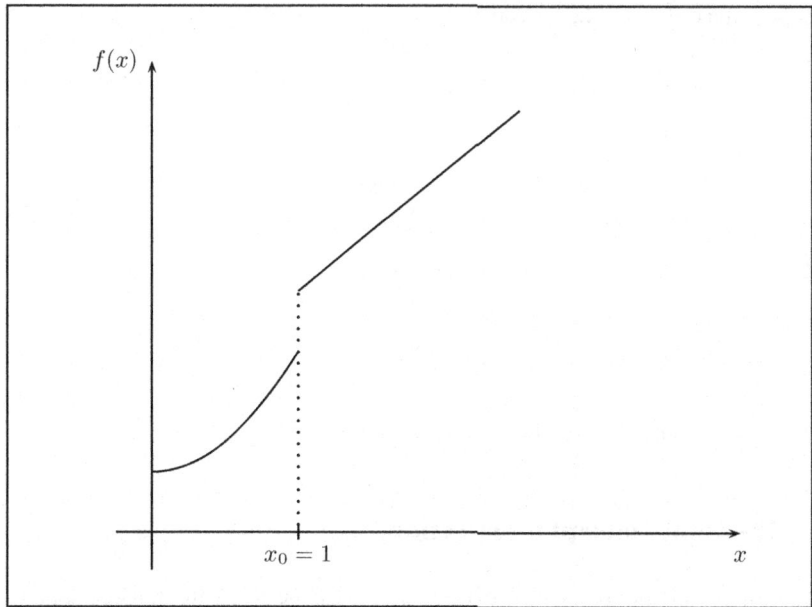

Fig. 2.29 To left- and right-side limit

If one considers the sectionally defined function

$$f(x) = \begin{cases} 2x^2 + 1 & \text{for } x \leq 1 \\ 3x + 1 & \text{for } x > 1, \end{cases}$$

then the join $x_0 = 1$ must be examined in particular.

The limit of this function at the point $x_0 = 1$ depends on whether one approaches from the right or from the left of the point $x_0 = 1$ (Fig. 2.29).

It applies:

$$\lim_{x \to x_0 + 0} f(x) = 4 \quad \text{and} \quad \lim_{x \to x_0 - 0} f(x) = 3.$$

Note
The notation

$$\lim_{x \to x_0 + 0} f(x) \quad \text{resp.} \quad \lim_{x \to x_0 - 0} f(x)$$

indicates that one approaches the point $x_0 = 1$ from the right $(x \to x_0 + 0)$, or from the left $(x \to x_0 - 0)$ side.

We therefore have to distinguish between a left- and right-side limit at a point x_0. Only if these two limits agree, do we speak of the existence of the limit at x_0.

2.6 Limits

▶ **Definition (Limit)** The limit $\lim_{x\to x_0} f(x)$ exists and has the value g, if the right -and left-side limit coincide:

$$\lim_{x\to x_0+0} f(x) = \lim_{x\to x_0-0} f(x) = g.$$

To use the coincidence of the right-and left-side limit as a criterion for the existence of the limit is a very pragmatic approach that is of great use for the continuity and differentiability investigations of elementary functions and sectionally defined functions in the following chapters. A mathematically deeper definition of the existence of a limit is the criterion named after the French mathematician Augustin Cauchy (1789–1857), which is introduced in the next section for the interested reader, but is not used further in the course of the book.

In addition to the sectionally defined functions, there are also other functions for which the consideration of limits $x \to x_0$ is interesting. An example is the hyperbolic function (see Sect. 2.2.6)

$$f(x) = \frac{1}{x}.$$

For this function, the point $x_0 = 0$ is particularly interesting, at which the denominator becomes zero. It holds:

$$\lim_{x\to 0+0} \frac{1}{x} = \infty \quad \text{and} \quad \lim_{x\to 0-0} \frac{1}{x} = -\infty.$$

This means, if you approach the critical value $x_0 = 0$ from the right, then $f(x) \to \infty$. If you approach from the left, then $f(x) \to -\infty$. Although ∞ is not a number and the equality sign is not correct, this way of writing is often chosen to express that the function values grow beyond all limits.

The point x_0 is called the **pole** of the function $f(x)$ and it is said that the function $f(x) = x^{-1}$ is *divergent* at the point $x_0 = 0$. Poles play an important role in fractional rational functions, they are treated in more detail there (see Sect. 2.6.6).

The concept of the limit is also used to investigate the behavior of functions for arbitrarily large or small x. This is expressed in the following way:

$$\lim_{x\to\infty} f(x) = g$$

or

$$\lim_{x\to-\infty} f(x) = g.$$

Examples for this are:

$$\lim_{x\to\infty} e^{-x} = 0$$

or

$$\lim_{x\to-\infty} e^{-x} = \infty.$$

Remark:
In general, the behavior of a function $f(x)$ for $x \to \pm\infty$ is also called its asymptotic behavoir.

2.6.3 Cauchy's Definition of the Limit of Functions

To define the limit concept introduced in the previous section more precisely, Cauchy introduced the following definition for the existence of a limit of a function:[11]

▶ **Definition (Limit According to Cauchy)** The limit $\lim_{x \to x_0} f(x)$ exists and has the value g, if for all $\varepsilon > 0$ there is a number $\delta > 0$ such that:

$$|f(x) - g| < \epsilon \quad \text{for } |x - x_0| < \delta. \tag{2.23}$$

Intuitively, this criterion means that for any stripe, no matter how small, of width ε, a strip of width δ can be found from which the x values are taken so that the distance of the function values from the limit is less than ε, see Fig. 2.30.

Example Let

$$f(x) = x^2.$$

We investigate with the help of the Cauchy criterion whether the limit

$$\lim_{x \to 0} f(x) = \lim_{x \to 0} x^2$$

exists. For this purpose, we must find a suitable δ, so that condition Eq. (2.23) is fulfilled for each ε.

$$|f(x) - g| < \epsilon \quad \text{for} \quad |x - x_0| < \delta$$
$$|x^2 - 0| < \epsilon \quad \text{for} \quad |x - 0| < \delta$$
$$\Longleftrightarrow |x^2| < \epsilon \quad \text{for} \quad |x| < \delta$$
$$\Longleftrightarrow |x| < \sqrt{\epsilon} \quad \text{for} \quad |x| < \delta.$$

We choose $\delta < \sqrt{\epsilon}$ here, then the condition is fulfilled for all $\varepsilon > 0$.

Next, we consider the sectionally defined function

$$f(x) = \begin{cases} 2x^2 + 1 & \text{for } x \leq 1 \\ 3x + 1 & \text{for } x > 1. \end{cases}$$

---

[11] A detailed discussion of these aspects can be found, for example, in Spivak's book (2008).

2.6 Limits

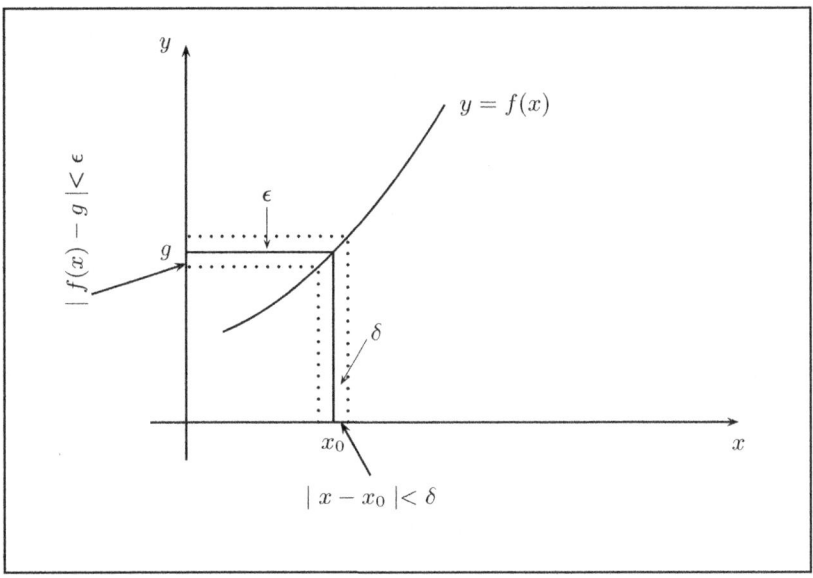

Fig. 2.30 To the Cauchy criterion

Using the Cauchy criterion, we now show that $g = 3$ is not the limit for $x \to 1$. Consider $x > 1$, then:

$$|f(x) - g| < \epsilon \quad \text{for} \quad |x - x_0| < \delta$$
$$|3x + 1 - 3| < \epsilon \quad \text{for} \quad |x - 1| < \delta$$
$$\stackrel{x \geq 1}{\Longleftrightarrow} \quad 3x + 1 - 3 < \epsilon \quad \text{for} \quad x - 1 < \delta$$
$$\Longleftrightarrow \quad x < \frac{1}{3}(\epsilon + 2) \quad \text{for} \quad x < 1 + \delta.$$

For each $\varepsilon > 0$, there must be a $\delta > 0$, so that this statement is fulfilled. Because $\delta > 0$, this is not given for $\varepsilon < 1$. Thus, $g = 3$ is not the limit of the function $f(x)$ for $x \to 1$.

2.6.4 Limit Considerations of Some Elementary Functions

In this section, we look at the limits of some elementary functions.

1. **Constant**

    A constant function

    $$f(x) = c; \quad c \in \mathbb{R}$$

has the limit
$$\lim_{x \to \pm\infty} f(x) = c.$$

2. **Power functions**
   For the power function it holds:
   $$\lim_{x \to \infty} x^n = \infty; \quad n \in \mathbb{N}$$
   and
   $$\lim_{x \to -\infty} x^n = \begin{cases} \infty & \text{for } n \text{ even} \\ -\infty & \text{for } n \text{ odd} \end{cases} \tag{2.24}$$
   and
   $$\lim_{x \to 0} x^n = 0; \quad n \in \mathbb{N}.$$

3. **Negative powers**
   For negative powers one has the following limits:
   $$\lim_{x \to \pm\infty} x^{-n} = 0; \quad n \in \mathbb{N}. \tag{2.25}$$

4. **Exponential function**
   The exponential function
   $$f(x) = a^x$$
   has the following asymptotic behavior:
   $$\lim_{x \to \infty} a^x = \begin{cases} 0 & \text{if } 0 < a < 1 \\ \infty & \text{if } a > 1. \end{cases}$$
   For $a > 1$, a^x grows faster than any power of x:
   $$\lim_{x \to \infty} \frac{x^n}{a^x} = 0 \quad \text{for} \quad n > 0, a > 1.$$
   An explanation for this is given in Sect. 3.6.1.

5. **Logarithm function**
   The logarithm function has the following limit behavior:
   $$\lim_{x \to \infty} \log_a x = \infty \tag{2.26}$$

2.6 Limits

and

$$\lim_{x \to 0+0} \log_a x = -\infty. \qquad (2.27)$$

6. Other interesting limits are:

$$\lim_{x \to \infty} \left(1 + \frac{1}{x}\right)^x = e \qquad (2.28)$$

and

$$\lim_{x \to 0} (1 + x)^{\frac{1}{x}} = e, \qquad (2.29)$$

where e is the Euler number with the numerical value $e \approx 2,7182....$[12]

2.6.5 Calculation Rules for Limits

The limit theorems of sequences (see Eqs. (2.9a–2.9d)) can be used to prove the limit theorems for functions.

Let the following limits be given:

$$\lim_{x \to x_0} f_1(x) = g_1$$

$$\lim_{x \to x_0} f_2(x) = g_2.$$

Then the following applies to the:

1. Addition

$$\lim_{x \to x_0} (f_1(x) + f_2(x)) = g_1 + g_2. \qquad (2.30)$$

2. Multiplication

$$\lim_{x \to x_0} (f_1(x) \cdot f_2(x)) = g_1 \cdot g_2. \qquad (2.31)$$

3. Division

$$\lim_{x \to x_0} \frac{f_1(x)}{f_2(x)} = \frac{g_1}{g_2}, \quad \text{für } g_2 \neq 0. \qquad (2.32)$$

---

[12] A rigorous derivation of this limit can be found in Courant and Robbins (2000), Chap. VI.3 or in Maor (2015), Appendix 2.

In addition, for chained functions

$$\lim_{x \to x_0} f(g(x)) = f\left(\lim_{x \to x_0} g(x)\right). \tag{2.33}$$

With the calculation rules (2.30) to (2.33) and some basic limits, many limits of composite functions can be easily determined. In the next section we will look at some examples of this.

2.6.6 Examples of Limit Considerations

1. **Polynominals**
   First, we consider limits of Polynomials:

$$f(x) = \sum_{i=0}^{n} a_i x^i.$$

   For the limit behavior when x goes to $\pm\infty$, only the highest power of x is decisive. All smaller powers of x no longer play a role when x is large or small enough. The sign of the cofficient a_n determines, whether the values of the poynomial go to $+\infty$ or $-\infty$ when x goes to $\pm\infty$.

2. **Fractional rational functions**
   For the fractional rational functions

$$f(x) = \frac{a_n x^n + a_{n-1} x^{n-1} + \ldots + a_2 x^2 + a_1 x^1 + a_0}{b_m x^m + b_{m-1} x^{m-1} + \ldots + b_2 x^2 + b_1 x^1 + b_0}$$

   the limit behavior for x going to $\pm\infty$ is of interest, but also the behavior of the function when approaching a gap in the domain of definition—that is, a zero of the denominator. First, we consider the behavior for x going to $\pm\infty$. To obtain the limit of a fractional rational function, one brackets the highest power of x that occurs in the denominator in the numerator and denominator and then applies the limit rules. The following cases must be distinguished:

   (a) Numerator power = Denominator power ($n = m$)
      The function converges to the value: $\frac{a_n}{b_n}$.
   (b) Numerator power < Denominator power ($n < m$)
      The function has the limit zero.
   (c) Numerator power > Denominator power ($n > m$)
      The function diverges, i.e. it goes to $+\infty$ or $-\infty$.

2.6 Limits

Examples Let

$$f(x) = \frac{x^4 + 3x^2 + 1}{2x^4 - \frac{1}{2}x^3} \qquad (2.34)$$

First, it is not clear which limit the function (2.34) for $x \to \infty$ has, because:

$$\lim_{x \to \infty} f(x) = \frac{\infty}{\infty}.$$

By factoring out x^4 in the numerator and denominator, the following transformations can be made:

$$\begin{aligned}
\lim_{x \to \infty} f(x) &= \lim_{x \to \infty} \frac{x^4 + 3x^2 + 1}{2x^4 - \frac{1}{2}x^3} \\
&= \lim_{x \to \infty} \frac{x^4(1 + 3 \cdot \frac{1}{x^2} + \frac{1}{x^4})}{x^4(2 - \frac{1}{2x})} \\
&= \lim_{x \to \infty} \frac{1 + 3\frac{1}{x^2} + \frac{1}{x^4}}{2 - \frac{1}{2x}} \\
&\stackrel{(2.32)}{=} \frac{\lim_{x \to \infty}(1 + 3\frac{1}{x^2} + \frac{1}{x^4})}{\lim_{x \to \infty}(2 - \frac{1}{2x})} \\
&\stackrel{(2.30)}{=} \frac{\lim_{x \to \infty} 1 + \lim_{x \to \infty} 3\frac{1}{x^2} + \lim_{x \to \infty}\frac{1}{x^4}}{\lim_{x \to \infty} 2 - \lim_{x \to \infty}\frac{1}{2x}} \\
&\stackrel{(2.25)}{=} \frac{1}{2}.
\end{aligned}$$

Here is an example in which the numerator power is greater than the denominator power:

$$f(x) = \frac{-2x^4 + x^2 - x}{3x^3 - 2x - 1}. \qquad (2.35)$$

It also applies to the function (2.35):

$$\lim_{x \to \infty} f(x) = \frac{\infty}{\infty}.$$

Now factor out x^3 in the numerator and denominator, and we will again make use of the limit laws for limits without explicitly mentioning this:

$$\begin{aligned}
\lim_{x \to \infty} f(x) &= \lim_{x \to \infty} \frac{-2x^4 + x^2 - x}{3x^3 - 2x - 1} \\
&= \lim_{x \to \infty} \frac{x^3(-2x + \frac{1}{x} - \frac{1}{x^2})}{x^3(3 - 2\frac{1}{x^2} - \frac{1}{x^3})} \\
&= \lim_{x \to \infty} \frac{-2x + \frac{1}{x} - \frac{1}{x^2}}{3 - 2\frac{1}{x^2} - \frac{1}{x^3}} \\
&= \lim_{x \to \infty} \frac{-2x}{3} = -\infty.
\end{aligned}$$

Accordingly, for $x \to -\infty$

$$\lim_{x \to -\infty} f(x) = \lim_{x \to -\infty} \frac{-2x^4 + x^2 - x}{3x^3 - 2x - 1} = \lim_{x \to -\infty} \frac{-2x}{3} = +\infty.$$

Another way to investigate the asymptotics for fractional rational functions—that is, the behavior of $f(x)$ for $|x| \to \infty$—is the polynomial division.[13]

Example We consider the function (2.35)

$$f(x) = \frac{-2x^4 + x^2 - x}{3x^3 - 2x - 1}$$

and carry out a polynomial division:

$$\begin{array}{l}(-2x^4 + x^2 - x) : (3x^3 - 2x - 1) = -\frac{2}{3}x - \frac{\frac{1}{3}x^2 + \frac{5}{3}x}{3x^3 - 2x - 1} \\ \underline{-\left(-2x^4 + \frac{4}{3}x^2 + \frac{2}{3}x\right)} \\ -\frac{1}{3}x^2 - \frac{5}{3}x\end{array}$$

The fractional rational function $f(x)$ approaches for $|x| \to \infty$ the linear function $-\frac{2}{3}x$. The expression

$$\frac{\frac{1}{3}x^2 + \frac{5}{3}x}{3x^3 - 2x - 1}$$

goes to zero.

In the limit consideration, the limit values that lead to expressions of the form:

$$\frac{\infty}{\infty} \quad \text{or} \quad \frac{0}{0}$$

are usually always interesting. We will learn about other methods of determining such limit values with the help of differential calculus in Sect. 3.6.1.

Now let's look at the behavior of fractional rational functions when approaching gaps in the definition range, which are given by the zeros of the denominator. We already know the behavior from the hyperbolic function:

$$\lim_{x \to x_0 + 0} f(x) = \lim_{x \to 0+0} \frac{1}{x} = \infty \quad \text{and} \quad \lim_{x \to x_0 - 0} f(x) = \lim_{x \to 0-0} \frac{1}{x} = -\infty.$$

If a fractional rational function has a zero x_0 in the denominator, it can be brought into the form:

$$f(x) = \frac{(x - x_0)^n z(x)}{(x - x_0)^m n(x)}.$$

---

[13] See Sect. 2.2.5.

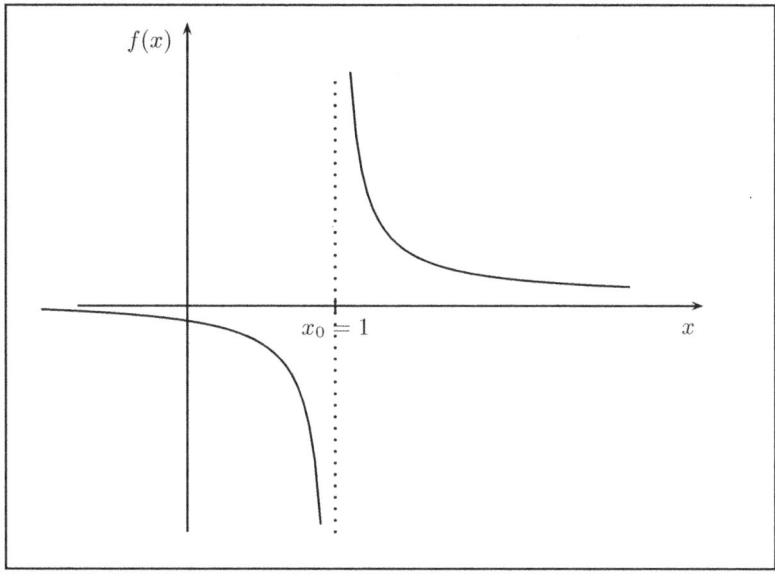

Fig. 2.31 Graph of the function $f(x) = (x + 2)/(x − 1)$ with a pole at $x_0 = 1$

Here $z(x)$ and $n(x)$ are polynominals that are not equal to zero in x_0, furthermore $m \geq 1$ and $n \geq 0$.

For $m > n$ there is a pole at x_0. If $m - n$ is odd, there is a change of sign at this pole, if $m - n$ is even, there is a pole without a change of sign.

Example
$$f(x) = \frac{x^2 + x - 2}{(x-1)^2}; \quad x \neq 1$$
$$\iff f(x) = \frac{(x-1)(x+2)}{(x-1)^2}$$
$$\iff f(x) = \frac{x+2}{x-1}.$$

At the point $x_0 = 1$ there is a pole with a change of sign (Fig. 2.31).

If $m \leq n$, there is no pole, but a removable gap in the domain of definition. We will come back to this in connection with the concept of the continuity of functions.

In economics, growth functions play a big role. A function that describes inhibited growth that starts at a value f_0 and then strictly monotonically gradually saturates is called a **logistic function**:

$$f(x) = \frac{a}{1 + be^{-cx}}; \quad x \in \mathbb{R}^+$$

with the positive parameters $a, b, c \in \mathbb{R}^+$.

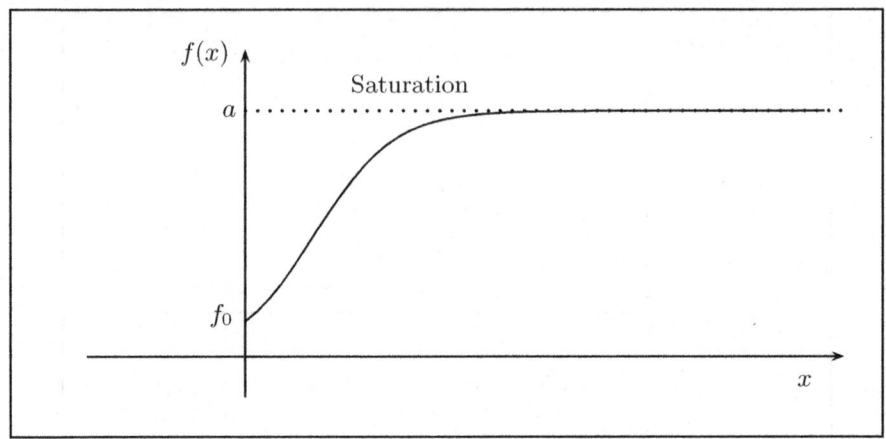

Fig. 2.32 The logistic function

The behavior of this function for $x \to 0$ and $x \to \infty$ can be determined by applying the limit theorems as follows (Fig. 2.32):

$$\lim_{x \to \infty} f(x) = \lim_{x \to \infty} \frac{a}{1 + b \cdot e^{-cx}}$$
$$= \frac{\lim_{x \to \infty} a}{\lim_{x \to \infty} (1 + b \cdot e^{-cx})}$$
$$= \frac{a}{1 + b \cdot \lim_{x \to \infty} e^{-cx}}$$
$$= \frac{a}{1}$$
$$= a.$$

This makes a the *saturation value* of the logistic function.

The behavior at $x = 0$ is:

$$\lim_{x \to 0} f(x) = \lim_{x \to 0} \frac{a}{1 + b \cdot e^{-cx}}$$
$$= \frac{a}{1 + b \cdot e^{-c0}}$$
$$= \frac{a}{1 + b}$$
$$= f_0.$$

2.7 Continuity of Functions

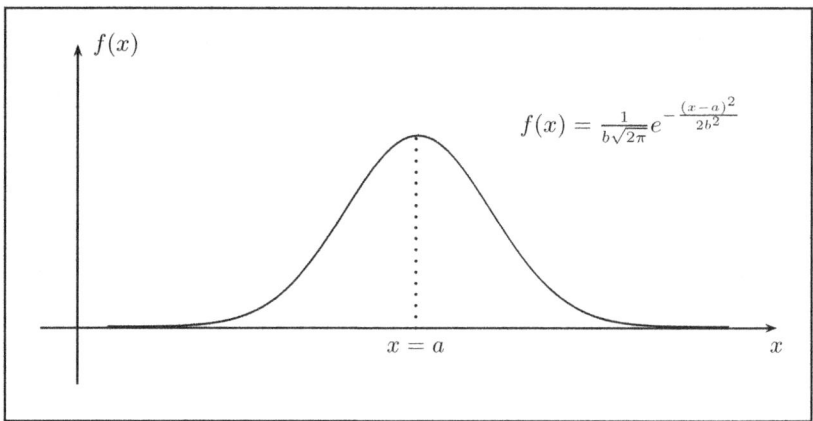

Fig. 2.33 The Gaussian distribution

In statistics and probability theory, the so-called *Gaussian distribution* plays a major role. This function is explicitly given by:

$$f(x) = \frac{1}{b\sqrt{2\pi}} e^{-\frac{(x-a)^2}{2b^2}}$$

with two real parameters a, b (see Fig. 2.33).

For this function, the following applies:

$$\lim_{x \to \pm\infty} f(x) = \lim_{x \to \pm\infty} \frac{1}{b\sqrt{2\pi}} e^{-\frac{(x-a)^2}{2b^2}} = 0.$$

2.7 Continuity of Functions

Under the continuity of a function, one understands, in an intuitive way, that the function graph can be traversed without taking a break, 'in one go'. A kink in the graph is therefore allowed, a jump is not. After this intuitive consideration, the cases shown in Fig. 2.34 are to be distinguished.

The two function graphs a and b on the left in Fig. 2.34 represent continuous functions. The graphs on the right c and d cannot be traversed 'in one go'. The function 2.34c is not continuous at the point x_0 because there is a jump there. In Fig. 2.34d continuity at the point x_0 cannot be spoken of because the function is not defined there. Continuity is a local property of a function, i.e. the property of continuity requires that the function is defined. The function 2.34d is continuous throughout the entire domain of definition, but has a gap in the domain of definition that cannot be closed so that the function is continuous in x_0.

Mathematically more precisely formulated:

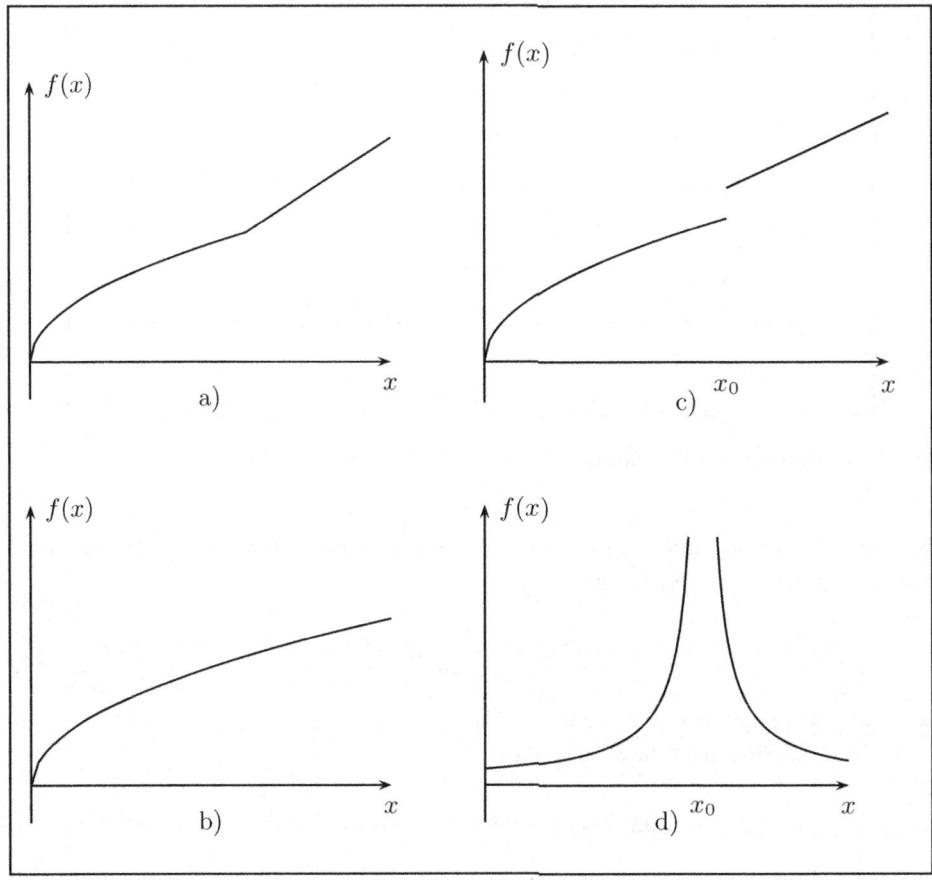

Fig. 2.34 To the concept of the continuity of a real function

▶ **Definition (Continuity)** A function $f(x)$ is continuous at the point x_0, if the limit at the point x_0 exists and coincides with the function value in x_0.

$$\lim_{x \to x_0+0} f(x) = \lim_{x \to x_0-0} f(x) = f(x_0). \tag{2.36}$$

In this definition we have made use of the definition of the existence of a limit from Sect. 2.6.2.

With this definition of continuity we take up the examples from Fig. 2.34 again. Elementary functions are continuous in the domain of definition. A function that describes the graph 2.34b is, for example.:

$$f(x) = \sqrt{x} \quad x \in \mathbb{R}.$$

2.7 Continuity of Functions

For all $x \in \mathbb{R}$ this function is continuous. A function that is described by the graph 2.34a can be represented as sectionally defined function in the following way:

$$f(x) = \begin{cases} \sqrt{x} & \text{for } x \leq 1 \\ x & \text{for } x > 1. \end{cases} \quad (2.37)$$

For the function (2.37) it holds:

$$\lim_{x \to 1+0} f(x) = 1 \quad \lim_{x \to 1-0} f(x) = 1 \quad f(1) = 1.$$

Therefore, $f(x)$ in $x_0 = 1$ is continuous.

On the other hand, the function in the upper right (Fig. 2.34c) is not continuous. It can be represented by:

$$f(x) = \begin{cases} \sqrt{x} & \text{for } x \leq 1 \\ x+1 & \text{for } x > 1. \end{cases} \quad (2.38)$$

For the function (2.38) it holds:

$$\lim_{x \to 1+0} f(x) = 2 \quad \lim_{x \to 1-0} f(x) = 1 \quad f(1) = 1.$$

Thus it is not continuous in $x_0 = 1$. Finally, we consider the last example from Fig. 2.34d right below. A function of the following form describes the course of the curve:

$$f(x) = \frac{1}{(x-1)^2}; \quad x \in \mathbb{R} \setminus \{1\}. \quad (2.39)$$

The function (2.39) is not defined in $x_0 = 1$. The function values approach ∞ if one approaches the point $x_0 = 1$ from the right or left, $f(x = 1)$ is not defined.

Sometimes there is a gap in the definition range that can be remedied. For this we consider the function (Fig. 2.35)

$$f(x) = \frac{x^2 - 1}{x - 1}; \quad x \in \mathbb{R} \setminus \{1\}.$$

With

$$\frac{x^2 - 1}{x - 1} = \frac{(x-1) \cdot (x+1)}{x - 1} = x + 1$$

we can write $f(x)$ as:

$$f(x) = x + 1 \,; x \in \mathbb{R} \setminus \{1\}.$$

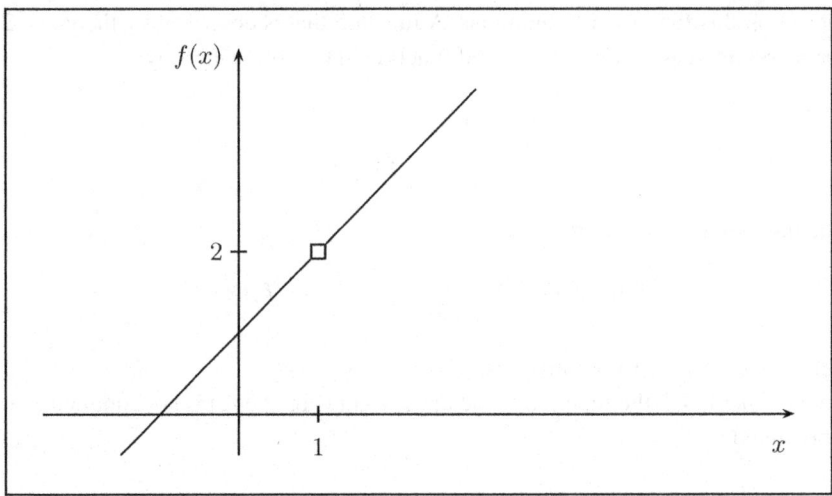

Fig. 2.35 The function $f(x) = \frac{x^2-1}{x-1}$ on $\mathbb{R} \setminus \{1\}$

For $f(x)$ the limits apply at the point $x_0 = 1$:

$$\lim_{x \to 1+0} f(x) = \lim_{x \to 1-0} f(x) = 2.$$

The value of the function at the point $x_0 = 1$, $f(1)$, is not defined.

An additional definition in the form $f(1) = 1$ is possible here. This can close the gap in the definition range so that the function is continuous.

2.8 Exercises

Short solutions to the following exercises can be found in the appendix.

2.1 Determine the maximum possible definition range of the functions

(a) $f(x) = \sqrt{x^2 - x - 2}$.

(b) $f(x) = \dfrac{1}{\ln(1 - x^2)}$.

2.2 Given the following functions:

$$f : \mathbb{R} \longrightarrow \mathbb{R}, \quad f(x) = x^2 - 5,$$
$$g : \mathbb{R} \setminus \{-2, +2\} \longrightarrow \mathbb{R}, \quad g(x) = \frac{5x}{x^2 - 2}.$$

2.8 Exercises

Form the concatenations

$$f \circ f, \quad f \circ g \quad \text{and} \quad g \circ f.$$

Calculate $f(g(1))$ and $g(f(1))$.

2.3 Determine the amplitude, circular frequency and phase shift as well as the zeros of the function

$$f(x) = 2\sin(3\pi x - 1.5\pi).$$

2.4 In the domain of definition $D_f = \{x \in \mathbb{R} \mid -1 \leq x \leq +2\}$, the assignment rule $y = 2x - 1$ applies. Which value range results?

2.5 For the general quadratic function

$$f(x) = ax^2 + bx + c$$

with $a \neq 0$, determine the maximum permissible definition and value range. What relationship must exist between the parameters a, b, c so that $W_f = \{y \mid y \geq 1\}$?

2.6 Draw the graphs of the following piecewise defined functions:

(a)
$$f(x) = \begin{cases} \frac{1}{2}x^2 - x & \text{for} \quad -3 \leq x \leq 0 \\ -\frac{1}{2}x^2 + x & \text{for} \quad 0 < x \leq 3. \end{cases}$$

(b)
$$f(x) = \begin{cases} \frac{1}{x} & \text{for} \quad 1 \leq x \leq 2 \\ \frac{1}{2}\sqrt{x-1} & \text{for} \quad x > 2. \end{cases}$$

2.7 An annual percentage economic growth relative to the previous year is exponential growth. Determine k for the function

$$f(x) = f(0) \cdot e^{kx},$$

if the growth is 2.5% per year. In what time period does the economic output double?

2.8 Show by determining the inverse mapping that the following functions define a one-to-one mapping of D_f onto W_f:

(a) $f(x) = \frac{1}{2}x + 3$ with $D_f = \{x \in \mathbb{R} \mid -10 \leq x \leq 10\}$.
(b) $f(x) = 4x^2 + 1$ with $D_f = \{x \in \mathbb{R} \mid x \geq 0\}$.

What is the domain of each inverse function? How can you tell the one-to-one mapping from the graph of the function?

2.9 Which of the following functions are limited upwards, which downwards? Which are limited?

(a) $f(x) = 5 - \frac{1}{x}$; $x > 0$.
(b) $f(x) = 7 + \frac{1}{x}$; $x < 0$.
(c) $f(x) = 2^x$; $-\infty < x < +\infty$.
(d) $f(x) = \dfrac{2x}{x+1}$; $x \geq 0$.

2.10 Consider the function

$$f : [-4, +4] \longrightarrow [-3, 8]$$

with

$$f(x) = \frac{1}{2}x + 4.$$

Investigate whether this function is injective and surjective.

2.11 Examine the following functions for injectivity, surjectivity and bijectivity:

(a) $f : \mathbb{R} \longrightarrow \mathbb{R}$; $x \longmapsto f(x) = x^2$
(b) $f : \mathbb{R} \longrightarrow \mathbb{R}^+$; $x \longmapsto f(x) = x^2$
(c) $f : \mathbb{R}^+ \longrightarrow \mathbb{R}^+$; $x \longmapsto f(x) = x^2$
(d) $k : \mathbb{R}^+ \setminus \{0\} \longrightarrow \{2,0; 2,2; 2,5; 2,75; 2,95\}$ with:

$$k(x) = \begin{cases} 2{,}0 & \text{for } \ 0 < x \leq 5000 \\ 2{,}2 & \text{for } \ 5001 < x \leq 7500 \\ 2{,}5 & \text{for } \ 7501 < x \leq 10.000 \\ 2{,}75 & \text{for } 10.001 < x \leq 20.000 \\ 2{,}95 & \text{for } 20.001 < x. \end{cases}$$

2.12 Given is the sectionally defined function

$$f(x) = \begin{cases} a \cdot e^x & \text{for } x \geq 1, \quad a \in \mathbb{R} \\ 2x + 1 & \text{for } x < 1. \end{cases}$$

Determine the constant a so that the function $f(x)$ is continuous.

2.13 Determine the domain and zeros of the following functions:

(a) $f(x) = 3e^{-x} - e^{2x}$.
(b) $f(x) = \frac{1}{2}(e^x + e^{-x})$.
(c) $f(x) = \frac{1}{2}(e^x - e^{-x})$.
(d) $f(x) = 3x^2 \cdot e^{-x^2} - 12e^{-x^2}$.
(e) $f(x) = 7 \cdot e^{\{\frac{x-1}{x+3}\}}$.

2.14 Determine the domain, zeros, and inverse function of the following functions:

(a) $f(x) = \ln \sqrt{x^2 + 1}$.
(b) $g(l) = \ln \frac{l}{2}$.
(c) $f(x) = \ln(x+1) + \ln x$.
(d) $h(b) = \ln b + \ln \sqrt{b^2 - 1}$.

2.15 Show that the function

$$f(x) = \frac{x}{e^x - 1} + \frac{x}{2}$$

is an even function.

2.16 Determine the general solution of the equation

$$A \sin[b(x+c)] = d, \quad a, b, c, d \in \mathbb{R}$$

2.17 Determine the parameters A, b, c, d of a sine function

$$f(x) = A \cdot \sin[b(x+c)] + d,$$

that describes periodic sales fluctuations over a year. Sales fluctuate between 0 and the maximum value U_{max}. On April 1, sales were $\frac{1}{2}U_{max}$ and then increased. The variable x describes the months, assume that all months are of equal length.

2.18 Investigate the two series

(a) the harmonic series $\sum_{n=1}^{\infty} \frac{1}{n}$,

(b) the series $\sum_{n=1}^{\infty} \frac{1}{n^2}$

with respect to convergence using the quotient criterion.

2.19 Determine the radius of convergence of each of the following power series:

$$p_1(x) = \sum_{n=0}^{\infty} \frac{x^n}{10^n},$$

$$p_2(x) = \sum_{n=0}^{\infty} n^2 x^n,$$

$$p_3(x) = \sum_{n=0}^{\infty} \frac{n!}{(2n)!} x^n,$$

$$p_4(x) = \sum_{n=0}^{\infty} \frac{n!}{n^n} \cdot x^n.$$

2.20 Examine whether the following limits exist and, if so, determine them:

(a) $\lim_{x \to -\infty} \dfrac{5x^2 + 7x - 1}{-3x^3 + 5x^2 + 25}$

(b) $\lim_{x \to 0} \dfrac{\frac{2}{x} + \frac{8}{x^2}}{4 + \frac{25}{x^2} + \frac{9}{x^3}}$

(c) $\lim_{x \to 0} \dfrac{\frac{2}{x^2} + \frac{5}{x}}{\frac{5}{x}}$

(d) $\lim_{x \to 3} \dfrac{3x - 9}{|12 - 4x|}$

2.21 Given the linear cost function

$$K(x) = K_F + c \cdot x.$$

Determine the fixed costs K_F and the constant c, if for 100 pieces production costs of 2000 € arise and for 300 pieces costs of 4000 €.

2.22 The logistic function

$$f(t) = \frac{32}{1 + 4 \cdot e^{-t/2}}$$

describes a saturation process.

(a) Calculate the saturation level.
(b) Determine the time points at which 50%, 70% and 90% of the saturation level are reached. Do these depend on the height of the saturation?

2.23 Determine the inverse function $p(x_N)$ belonging to the monotonically decreasing demand function

$$x_N(p) = x_0 - cp$$

Is this function also monotonically decreasing?

2.24 Determine the market equilibrium for the demand function

$$x_N(p) = x_0 - cp; \quad x_0, c > 0$$

and the supply function

$$x_A(p) = a_0 + bp^2; \quad a_0, b > 0.$$

Under what condition does a market equilibrium occur?

2.25 Determine all solutions of the following equations with the given solutions.

$x^3 + 6x^2 + 11x + 6 = 0, x_1 = -1.$
$x^4 + 5x^3 - 19x^2 - 65x + 150 = 0, x_1 = 2, x_2 = -5.$

2.26 Determine the following limits.

(a) $\lim_{x \to 1} \dfrac{x^2 - 1}{x - 1}$,

(b) $\lim_{x \to -1/2} \dfrac{4x^2 - 1}{2x + 1}$,

(c) $\lim_{x \to 1} \dfrac{1 - x}{1 - \sqrt{x}}$,

References

Arens T., Hettlich F., Karpfinger Ch., Kockelkorn U., Lichtenegger K., Stachel H. (2018): Mathematik, 4. Auflage, Spektrum Akademischer Verlag, Heidelberg.
Courant R., Robbins H. (2000): Was ist Mathematik? Fünfte, unveränderte Auflage, Springer, Berlin, Heidelberg, New York.
Maor E. (2017): To Infinity and Beyond, New Edition, Princeton University Press, Princeton, New Jersey.
Maor E. (2015): e: The Story of a Number, Princeton University Press, Princeton, New Jersey.
Spivak M. (2008): Calculus, Third Edition, Cambridge University Press, Cambridge.

Differential Calculus 3

> **Learning Objectives (This Chapter Covers)**
>
> - how the derivative of a function is defined
> - under which conditions the derivative is defined
> - the application of differential calculus in the context of discussion of a curve
> - the determination of zeros of a function using the Newton method
> - the application of differential calculus in economic questions
> - the approximation of functions by Taylor series ◀

3.1 The Concept of the Derivative

Motivation

Often, not only the function value at a point x_0 is of interest, but also the change in the function course in the vicinity of the point x_0.

In Fig. 3.1 the course of the profit over time for two companies U_1 and U_2 is sketched. The company U_1 has at the current time t_0 a higher profit than the company U_2. Nevertheless, one recognizes on the basis of the course of the curve that the profit of the second company grows and the profit of the first company decreases.

Changes are described by the gradient of a curve at a certain point x_0.

Slope of a Line

First, let's look at the slope of a straight line. The general form of an equation of a straight line is:

$$y = f(x) = mx + b$$

© The Author(s), under exclusive license to Springer-Verlag GmbH, DE, part of Springer Nature 2023
T. Holey and A. Wiedemann, *Analysis and Linear Algebra*,
https://doi.org/10.1007/978-3-662-66247-2_3

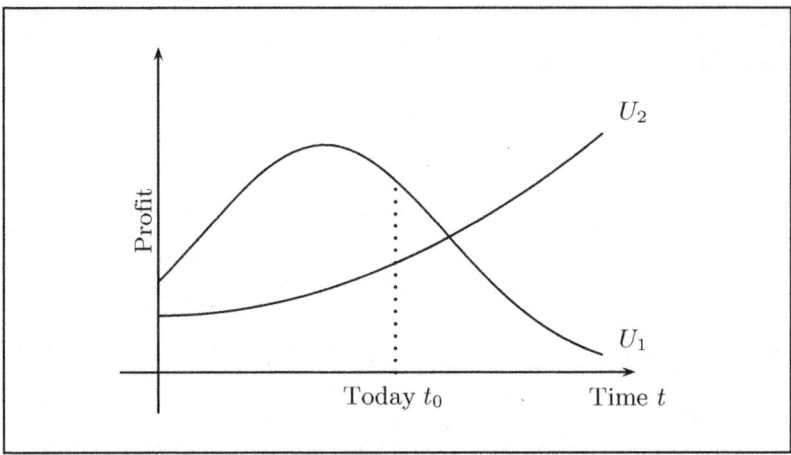

Fig. 3.1 In which company would one seek long-term employment?

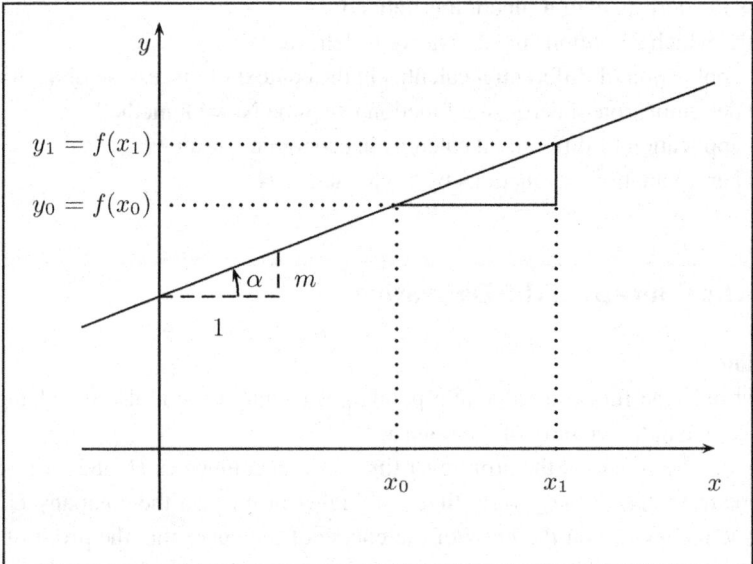

Fig. 3.2 The slope of a line

with the slope m and the intercept b. The slope of the line is the ratio of the change in y-direction to the change in x-direction. This is illustrated in Fig. 3.2.
For the slope of the line it holds:

$$m = \tan \alpha = \frac{m}{1} = \frac{y_1 - y_0}{x_1 - x_0} = \frac{f(x_1) - f(x_0)}{x_1 - x_0} = \frac{\Delta y}{\Delta x}.$$

3.1 The Concept of the Derivative

With

$$x_1 = x_0 + \Delta x$$

the slope of a line can be formulated as follows:

$$m = \frac{f(x_0 + \Delta x) - f(x_0)}{x_0 + \Delta x - x_0} = \frac{f(x_0 + \Delta x) - f(x_0)}{\Delta x}. \tag{3.1}$$

The quotient $\frac{f(x_0+\Delta x)-f(x_0)}{\Delta x}$ is called the **difference quotient** of $f(x)$ in the vicinity of x_0.

The slope of any function $f(x)$ at the point x_0 can be related to a the slope of a straight line using the following definition.

▶ **Definition (Slope of a Function)**
The slope of a function $f(x)$ at point x_0 is the slope of the tangent to $f(x)$ at point x_0.

The question is of course how to determine the tangent at a point. To do this, we make use of the concept of limit.

We start from a *secant* that intersects the function $f(x)$ in two points, $P_0 = (x_0, f(x_0))$ and $P_1 = (x_0 + \Delta x, f(x_0 + \Delta x))$ (see Fig. 3.3). As follows from the above considerations, the slope of the secant is:

$$m_s = \frac{f(x_0 + \Delta x) - f(x_0)}{\Delta x}.$$

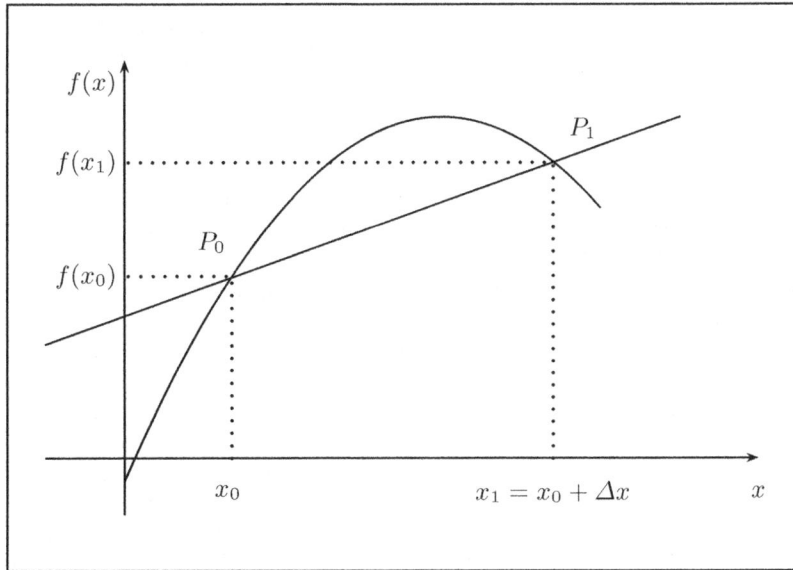

Abb. 3.3 To the slope of a function at a point x_0

We understand the limit transition as the tangent when $x_1 \to x_0$ holds, which is equivalent to $\Delta x \to 0$.

Therefore, we formulate the slope of the tangent as the limit of the secant slope:

$$m_t(x_0) = \lim_{\Delta x \to 0} m_s = \lim_{\Delta x \to 0} \frac{f(x_0 + \Delta x) - f(x_0)}{\Delta x} = \lim_{\Delta x \to 0} \frac{\Delta f}{\Delta x}.$$

$m_t(x_0)$ is the slope of the function $f(x)$ at the point x_0. It is also denoted by $f'(x_0)$. Two notations for the slope of a function at a point have become established.

▶ **Definition (Derivative of a Function)**

The derivative of a function $f(x)$ at the point $x_0 \in D_f$ is defined as:

$$m_t(x_0) = f'(x_0) = \lim_{\Delta x \to 0} \frac{f(x_0 + \Delta x) - f(x_0)}{\Delta x} \tag{3.2}$$

and

$$f'(x_0) = \frac{df(x)}{dx}\bigg|_{x=x_0}. \tag{3.3}$$

Read: 'df by dx for $x = x_0$' or 'the derivative of f with respect to x'.

The derivative of a function $f(x)$ is therefore itself a function of x. In this way, a slope value is assigned to each value of the domain of definition, and the **derivative function**

$$f'(x) = \frac{df(x)}{dx}$$

of $f(x)$ is created.[1] Instead of derivative function one also says **differential quotient**. This indicates that the derivative is the limit of a quotient, namely the **difference quotient**. We will address the question of the existence of this derivative function in Sect. 3.4.[2]

---

[1] Different notation is used in mathematical literature for the differentiation of a function. The designation $\frac{df(x)}{dx}$ goes back to Leibniz, the notation $f'(x)$ was introduced by Joseph Louis Lagrangre 1797. See Maor (2015), pp. 95 ff. The originally introduced by Newton notation $\dot{y}$ for the derivative is occasionally used in physics for the designation of the time derivative.

[2] The basic concepts of differential and integral calculus were developed independently of each other by Isaac Newton (1643–1727) in Cambridge, England and Gottfried Wilhelm Leibniz (1646–1716). Newton's focus was on the investigation of laws of motion—here, velocity is the derivative of a distance-time function with respect to time—while Leibniz was interested in more formal aspects.

There was a priority dispute between Newton and Leibniz from the beginning. Newton developed the calculus of differential in 1669, but did not publish his work until 1711, and a complete version was not published until 1736. Leibniz, on the other hand, developed the calculus of differential in 1676 and quickly disseminated his work on the continent. However, since Newton's work was informally disseminated among mathematicians from the beginning, Leibniz was accused of plagiarism by the English side. Detailed discussions of this classical priority dispute in the history of mathematics can be found in Alten et al. (2014), Dunham (1990), Hall (2002), Maor (2015) or Stillwell (2002).

3.2 Derivatives of Elementary Functions

In this section we examine the derivatives of some elementary functions with the definition from Sect. 3.1.

1. The derivative of the quadratic function:

$$f(x) = x^2.$$

With Eq. 3.2 it follows:

$$\begin{aligned} f'(x) &= \lim_{\Delta x \to 0} \frac{f(x + \Delta x) - f(x)}{\Delta x} \\ &= \lim_{\Delta x \to 0} \frac{(x + \Delta x)^2 - x^2}{\Delta x} \\ &= \lim_{\Delta x \to 0} \frac{x^2 + 2x\Delta x + (\Delta x)^2 - x^2}{\Delta x} \\ &= \lim_{\Delta x \to 0} \frac{\Delta x \cdot (2x + \Delta x)}{\Delta x} \\ &= \lim_{\Delta x \to 0} 2x + \Delta x = 2x. \end{aligned}$$

The slope of a parabola is therefore not a constant, but depends on the considered point x. This is illustrated again in Fig. 3.4. The derivative defined as a differential quotient therefore represents a function of x again.

2. The derivative of the hyperbolic function:

$$f(x) = \frac{1}{x}.$$

From Eq. (3.2) it follows:

$$\begin{aligned} f'(x) &= \lim_{\Delta x \to 0} \frac{f(x + \Delta x) - f(x)}{\Delta x} \\ &= \lim_{\Delta x \to 0} \frac{\frac{1}{x + \Delta x} - \frac{1}{x}}{\Delta x} \\ &= \lim_{\Delta x \to 0} \frac{\frac{x - (x + \Delta x)}{x(x + \Delta x)}}{\Delta x} \\ &= \lim_{\Delta x \to 0} \frac{1}{\Delta x} \cdot \frac{-\Delta x}{x(x + \Delta x)} \\ &= \lim_{\Delta x \to 0} \frac{-1}{x^2 + x \cdot \Delta x} \\ &= -\frac{1}{x^2}. \end{aligned}$$

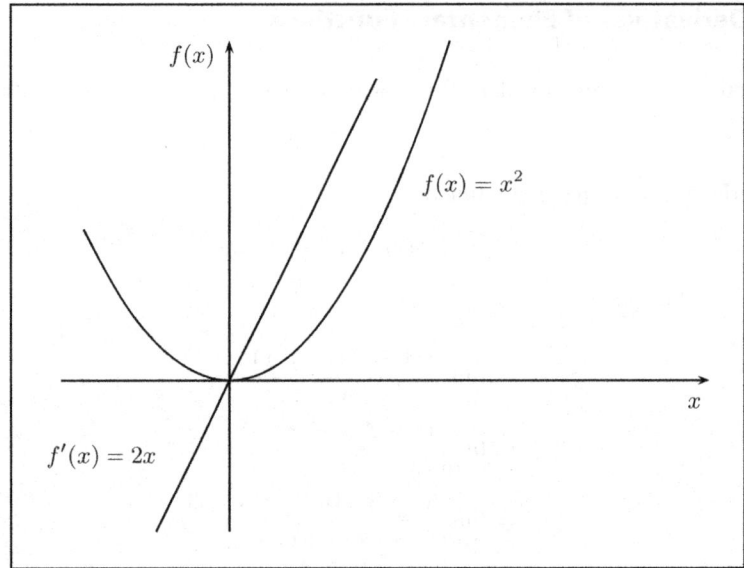

Fig. 3.4 The function $f(x) = x^2$ with its derivative function $f'(x) = 2x$

3. The derivative of the exponential function:

$$f(x) = e^x.$$

With Eq. (3.2) we get:

$$f'(x) = \lim_{\Delta x \to 0} \frac{f(x + \Delta x) - f(x)}{\Delta x}$$

$$= \lim_{\Delta x \to 0} \frac{e^{x+\Delta x} - e^x}{\Delta x}$$

$$= \lim_{\Delta x \to 0} e^x \cdot \frac{e^{\Delta x} - 1}{\Delta x}.$$

We now set:

$$e^{\Delta x} = h + 1 \quad \Longleftrightarrow \quad \Delta x = \ln(h + 1).$$

With $\Delta x \to 0$ we get $e^{\Delta x} \to 1$. Therefore we replace $\Delta x \to 0$ in the limit by $h \to 0$. This gives:

3.2 Derivatives of Elementary Functions

$$f'(x) = \lim_{h \to 0} e^x \cdot \frac{h}{\ln(1+h)}$$
$$= \lim_{h \to 0} e^x \cdot \frac{1}{\ln(1+h)^{1/h}}$$
$$= e^x \lim_{h \to 0} \frac{1}{\ln(1+h)^{1/h}}$$
$$= e^x \frac{1}{\ln(\lim_{h \to 0}(1+h)^{1/h})}$$
$$= e^x \frac{1}{\ln(e)}$$
$$= e^x.$$

We have used the rule (2.33):

$$\lim_{x \to x_0} f(g(x)) = f\left(\lim_{x \to x_0} g(x)\right).$$

Therefore we get:[3]

$$\boxed{f(x) = e^x; \qquad f'(x) = e^x.} \qquad (3.4)$$

The result (3.4) is: *The derivative of the exponential function is equal to the exponential function*. This means that the change of the exponential function—or the increase—is also exponential.

With similar considerations one can show:

$$\boxed{f(x) = \ln x; \quad (x > 0); \qquad f'(x) = \frac{1}{x}.} \qquad (3.5)$$

4. The derivative of the power function

For the derivative of the power function we have:

$$\boxed{f(x) = x^n; \qquad f'(x) = n \cdot x^{n-1}.} \qquad (3.6)$$

From Eq. (3.2) we get this by (compare also (Eq. 1.19)):

---

[3] See also Maor (2015), Chap. 10.

$$\begin{aligned}
\frac{df(x)}{dx} &= \lim_{\Delta x \to 0} \frac{f(x+\Delta x)-f(x)}{\Delta x} \\
&= \lim_{\Delta x \to 0} \frac{(x+\Delta x)^n - x^n}{\Delta x} \\
&= \lim_{\Delta x \to 0} \frac{\sum_{k=0}^{n} \binom{n}{k} x^{n-k}(\Delta x)^k - x^n}{\Delta x} \\
&= \lim_{\Delta x \to 0} \frac{1}{\Delta x}\left(x^n + \sum_{k=1}^{n} \binom{n}{k} x^{n-k}(\Delta x)^k - x^n \right) \\
&= \lim_{\Delta x \to 0} \frac{1}{\Delta x}\left(\sum_{k=1}^{n} \binom{n}{k} x^{n-k}(\Delta x)^k \right) \\
&= \lim_{\Delta x \to 0} \frac{1}{\Delta x}\left(n x^{n-1}(\Delta x)^1 + \sum_{k=2}^{n} \binom{n}{k} x^{n-k}(\Delta x)^k \right) \\
&= \lim_{\Delta x \to 0} \left(n x^{n-1} + \sum_{k=2}^{n} \binom{n}{k} x^{n-k}(\Delta x)^{k-1} \right) \\
&= n x^{n-1} + \lim_{\Delta x \to 0} \sum_{k=2}^{n} \binom{n}{k} x^{n-k}(\Delta x)^{k-1} \\
&= n x^{n-1}.
\end{aligned}$$

In the derivation we have assumed $n \in \mathbb{N}$. But the result can be extended to a larger range of n, which is given here without proof:

$$\begin{aligned}
n &\in \mathbb{N} \quad \text{and} \quad x \in \mathbb{R} \\
n &\in \mathbb{Z} \quad \text{and} \quad x \in \mathbb{R} \setminus \{0\} \\
n &\in \mathbb{R} \quad \text{and} \quad x \in \mathbb{R}^+.
\end{aligned}$$

3.3 Derivative Rules

In this chapter **derivative rules** are summarized, which follow from the definition of the derivative. These rules allow to calculate the derivative of sums, products, quotients or compositions of functions if the derivative of each factor is known.

1. **Constant factors and summands**
   A constant factor remains unchanged during differentiation, a constant summand is omitted:
   $$g(x) = cf(x) + d; \quad c, d \in \mathbb{R}.$$
   Then it applies:
   $$\boxed{g'(x) = cf'(x).} \tag{3.7}$$

3.3 Derivative Rules

2. Sums of functions
A sum of functions can be differentiated term by term. For
$$f(x) = u(x) + v(x)$$
it follows that
$$f'(x) = u'(x) + v'(x). \tag{3.8}$$

Proof

$$\begin{aligned}
f'(x) &= \lim_{\Delta x \to 0} \frac{f(x + \Delta x) - f(x)}{\Delta x} \\
&= \lim_{\Delta x \to 0} \frac{u(x + \Delta x) + v(x + \Delta x) - u(x) - v(x)}{\Delta x} \\
&= \lim_{\Delta x \to 0} \frac{u(x + \Delta x) - u(x) + (v(x + \Delta x) - v(x))}{\Delta x} \\
&= \lim_{\Delta x \to 0} \frac{u(x + \Delta x) - u(x)}{\Delta x} + \lim_{\Delta x \to 0} \frac{v(x + \Delta x) - v(x)}{\Delta x} \\
&= u'(x) + v'(x).
\end{aligned}$$

3. Product rule:
For the product of two functions:
$$f(x) = u(x) \cdot v(x)$$
the following rule applies:

$$\boxed{f'(x) = u'(x) \cdot v(x) + u(x) \cdot v'(x).} \tag{3.9}$$

Proof

$$\begin{aligned}
\frac{df(x)}{dx} &= \lim_{\Delta x \to 0} \frac{f(x + \Delta x) - f(x)}{\Delta x} \\
&= \lim_{\Delta x \to 0} \frac{u(x + \Delta x) \cdot v(x + \Delta x) - u(x) \cdot v(x)}{\Delta x} \\
&= \lim_{\Delta x \to 0} \frac{1}{\Delta x} \Big[u(x + \Delta x) \cdot v(x + \Delta x) - v(x + \Delta x) \cdot u(x) \\
&\qquad\qquad + v(x + \Delta x) \cdot u(x) - u(x) \cdot v(x) \Big] \\
&= \lim_{\Delta x \to 0} \frac{v(x + \Delta x) \cdot [u(x + \Delta x) - u(x)] + u(x) \cdot [v(x + \Delta x) - v(x)]}{\Delta x} \\
&= \lim_{\Delta x \to 0} \left[v(x + \Delta x) \cdot \frac{u(x + \Delta x) - u(x)}{\Delta x} + u(x) \frac{v(x + \Delta x) - v(x)}{\Delta x} \right] \\
&= \lim_{\Delta x \to 0} \left[v(x + \Delta x) \cdot \frac{u(x + \Delta x) - u(x)}{\Delta x} \right] \\
&\quad + \lim_{\Delta x \to 0} \left[u(x) \frac{v(x + \Delta x) - v(x)}{\Delta x} \right]
\end{aligned}$$

$$= \lim_{\Delta x \to 0} [v(x + \Delta x)] \cdot \lim_{\Delta x \to 0} \left[\frac{u(x + \Delta x) - u(x)}{\Delta x} \right]$$

$$+ u(x) \cdot \lim_{\Delta x \to 0} \left[\frac{v(x + \Delta x) - v(x)}{\Delta x} \right]$$

$$= v(x) \cdot \frac{du(x)}{dx} + u(x) \frac{dv(x)}{dx}.$$

From the rule (3.9) it follows in particular the derivation of the square of a function:

$$\frac{df(x)^2}{dx} = 2f(x) \cdot f'(x).$$

4. **Quotient rule:**

For the quotient of two functions

$$f(x) = \frac{u(x)}{v(x)}; \quad v(x) \neq 0$$

the rule applies:

$$\boxed{f'(x) = \frac{u'(x) \cdot v(x) - u(x) \cdot v'(x)}{[v(x)]^2}.} \quad (3.10)$$

Proof

$$\frac{df(x)}{dx} = \lim_{\Delta x \to 0} \frac{f(x + \Delta x) - f(x)}{\Delta x}$$

$$= \lim_{\Delta x \to 0} \frac{1}{\Delta x} \cdot \left\{ \frac{u(x + \Delta x)}{v(x + \Delta x)} - \frac{u(x)}{v(x)} \right\}$$

$$= \lim_{\Delta x \to 0} \frac{v(x) \cdot u(x + \Delta x) - u(x) \cdot v(x + \Delta x)}{v(x) \cdot v(x + \Delta x) \cdot \Delta x}$$

$$= \lim_{\Delta x \to 0} \frac{v(x) \cdot \frac{u(x+\Delta x)}{\Delta x} - u(x) \cdot \frac{v(x+\Delta x)}{\Delta x}}{v(x) \cdot v(x + \Delta x)}$$

$$= \lim_{\Delta x \to 0} \frac{v(x) \cdot \left(\frac{u(x+\Delta x)}{\Delta x} - \frac{u(x)}{\Delta x} \right) - u(x) \cdot \left(\frac{v(x+\Delta x)}{\Delta x} - \frac{v(x)}{\Delta x} \right)}{v(x) \cdot v(x + \Delta x)}$$

$$= \frac{v(x) \cdot \frac{du(x)}{dx} - u(x) \frac{dv(x)}{dx}}{v^2(x)}.$$

5. **Chain rule:**

For the composition of two functions

$$f(x) = f(g(x)); \quad \text{where } g = g(x)$$

3.3 Derivative Rules

it holds:

$$f'(x) = f'(g) \cdot g'(x), \tag{3.11}$$

or as differential quotient:

$$\frac{d}{dx} f(g(x)) = \frac{df(g(x))}{dg} \cdot \frac{dg(x)}{dx}. \tag{3.12}$$

Example Consider the function:

$$f(x) = (3x + 1)^2.$$

We set

$$g(x) = 3x + 1$$

and

$$f(g(x)) = \big[g(x)\big]^2.$$

Then according to the above chain rule Eq. (3.12):

$$f'(x) = 2 \cdot g(x) \cdot 3 = 2 \cdot (3x + 1) \cdot 3 = 18x + 6.$$

With the help of the chain rule we can also derive the following derivatives:

$$f(x) = a^x; \, a \in \mathbb{R}, a > 0.$$

Then

$$\frac{df(x)}{dx} = \ln a \cdot a^x. \tag{3.13}$$

Proof
We substitute:

$$a^x = \left(e^{\ln a}\right)^x = e^{x \ln a}.$$

$$\begin{aligned}
\frac{df(x)}{dx} &= \frac{d}{dx} a^x \\
&= \frac{d}{dx} (e^{\ln a})^x \\
&= \frac{d}{dx} (e^{x \cdot \ln a}).
\end{aligned}$$

If we set
$$g(x) = x \cdot \ln a,$$
then
$$\frac{df(x)}{dx} = \frac{d}{dg}e^g \cdot \frac{dg(x)}{dx}$$
$$= e^g \cdot \frac{d(x \ln a)}{dx}$$
$$= e^g \cdot \ln a$$
$$= e^{x \cdot \ln a} \cdot \ln a$$
$$= a^x \cdot \ln a.$$

The derivative of the logarithm function
$$f(x) = \log_a x; \quad a \in \mathbb{R}, a > 0$$
is:
$$\boxed{\frac{df(x)}{dx} = \frac{1}{\ln a} \cdot \frac{1}{x}.} \qquad (3.14)$$

Proof
We substitute
$$\log_a x = \frac{\ln x}{\ln a}.$$
So that:
$$\frac{d}{dx}\log_a x = \frac{d}{dx}\frac{\ln x}{\ln a}$$
$$= \frac{1}{\ln a}\frac{d \ln x}{dx} = \frac{1}{\ln a}\frac{1}{x}.$$

We summarize the derived derivative functions for the sake of clarity in Table 3.1.

3.4 Differentiability

So far, we have always assumed that the limit
$$\lim_{\Delta x \to 0} \frac{f(x + \Delta x) - f(x)}{\Delta x}$$
exists, that is, that the left and right hand side limits agree.

3.4 Differentiability

Table 3.1 Table of derivatives of some elementary functions

$f(x)$	$f'(x)$
$ax + b$	a
$\frac{1}{x}$	$-\frac{1}{x^2}$
$\sqrt{x}$	$\frac{1}{2\sqrt{x}}$
x^n	nx^{n-1}
e^x	e^x
$\ln x$	$\frac{1}{x}$
a^x	$\ln a \cdot a^x$
$\log_a x$	$\frac{1}{\ln a} \cdot \frac{1}{x}$
$\sin x$	$\cos x$
$\cos x$	$-\sin x$

To investigate this aspect in more detail, we consider the absolute value function

$$f(x) = |x| = \begin{cases} x & \text{for } x \geq 0 \\ -x & \text{for } x < 0. \end{cases}$$

We investigate the limit

$$\lim_{\Delta x \to 0} \frac{f(x_0 + \Delta x) - f(x_0)}{\Delta x}$$

for $x_0 = 0$ in more detail.

The right-hand limit is:

$$\lim_{\Delta x \to 0+0} \frac{f(0 + \Delta x) - f(0)}{\Delta x} = \lim_{\Delta x \to 0+0} \frac{0 + \Delta x - 0}{\Delta x} = 1.$$

The left-hand limit is:

$$\lim_{\Delta x \to 0-0} \frac{f(0 + \Delta x) - f(0)}{\Delta x} = \lim_{\Delta x \to 0-0} \frac{0 - \Delta x - 0}{\Delta x} = -1.$$

Since the left- and right-hand limits do not match, the limit does not exist. As can be seen from Fig. 3.5, the graph of the absolute value function has a kink at the point $x_0 = 0$. Such a kink represents a *non-differentiable point*.

Remark
The absolute value function $f(x) = |x|$ is continuous at the point $x_0 = 0$.
We prove this statement by applying the continuity criterion Eq. (2.36):

$$\lim_{x \to 0+0} f(x) = \lim_{x \to 0+0} |x| = \lim_{x \to 0+0} x = 0.$$

$$\lim_{x \to 0-0} f(x) = \lim_{x \to 0-0} |x| = \lim_{x \to 0-0} -x = 0 = f(0).$$

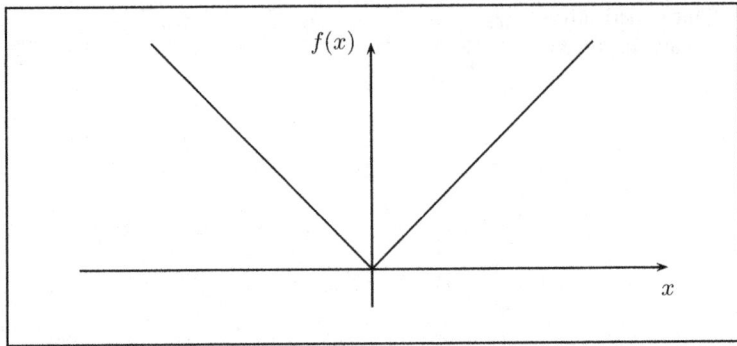

Fig. 3.5 The graph of the absolute value function $f(x) = |x|$

A function that is continuous at a point x_0 is therefore not necessarily differentiable at this point. However, if a function is not continuous at a point, it cannot be differentiable at that point.

We consider the following example (Fig. 3.6):

$$f(x) = \begin{cases} 2x & \text{for } x \leq 1 \\ 2x - 1 & \text{for } x > 1. \end{cases} \tag{3.15}$$

At first glance, this function appears to be differentiable at the point $x_0 = 1$, since the slope has the value 2 when approaching from the left or right of the point $x_0 = 1$.

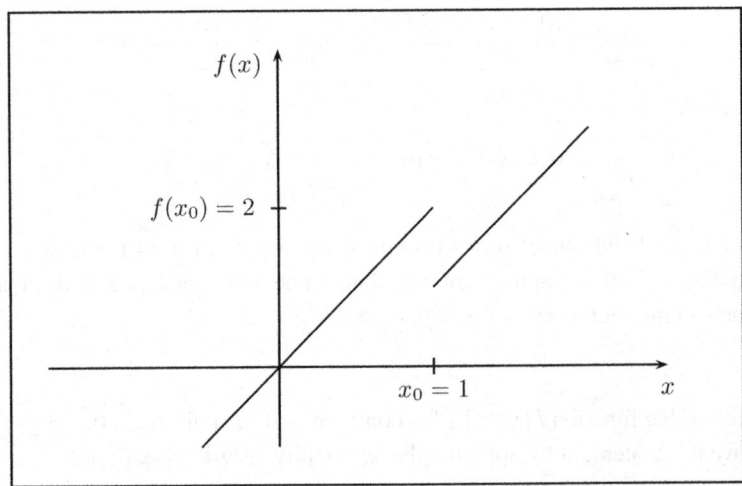

Fig. 3.6 The graph of the function (3.15)

However, the following considerations show that this assumption is false:

- Left-sided limit ($x < 1$), that is, with $x_0 = 1$ and $\Delta x < 0$:

$$\lim_{\Delta x \to 0} \frac{f(x_0 + \Delta x) - f(x_0)}{\Delta x} = \lim_{\Delta x \to 0} \frac{2(x_0 + \Delta x) - 2x_0}{\Delta x}$$
$$= \lim_{\Delta x \to 0} \frac{2x_0 + 2\Delta x - 2x_0}{\Delta x}$$
$$= \lim_{\Delta x \to 0} 2$$
$$= 2.$$

- Right-sided limit ($x > 1$), that is, with $x_0 = 1$ and $\Delta x > 0$:

$$\lim_{\Delta x \to 0} \frac{f(x_0 + \Delta x) - f(x_0)}{\Delta x} = \lim_{\Delta x \to 0} \frac{2(x_0 + \Delta x) - 1 - 2x_0}{\Delta x}$$
$$= \lim_{\Delta x \to 0} \frac{2\Delta x - 1}{\Delta x}$$
$$= \lim_{\Delta x \to 0} \left(2 - \frac{1}{\Delta x}\right)$$
$$= -\infty.$$

Thus, the function 3.15 is not differentiable, because the left- and right-sided limits do not agree.

The considerations from these two examples can be generalized to the following statement:

Continuity is a *necessary* condition for differentiability, continuity is not *sufficient* for differentiability.

Or, in other words: A function that is not continuous at a point x_0 is also not differentiable at that point. A function that is continuous at x_0 does not necessarily have to be differentiable at that point. On the other hand, we can conclude: If a function is differentiable at x_0, then it must also be continuous at that point.

3.5 Higher Derivatives, Extreme Values and Turning Points

The function $f'(x)$ is called the 1st derivative of the function $f(x)$. Since $f'(x)$ is again a function of x, higher derivatives can be formed:

$$f'(x) = \frac{df(x)}{dx}$$
$$f''(x) = \frac{d^2 f(x)}{dx^2}$$
$$f'''(x) = \frac{d^3 f(x)}{dx^3}$$
$$\vdots \qquad \vdots$$
$$f^{(n)}(x) = \frac{d^n f(x)}{dx^n}.$$

As we have seen, the first derivative characterizes the slope of a function $f(x)$. The 2nd derivative describes the change in slope (i.e. the slope of the slope). In the graph of a function, this can be interpreted as the curvature.

We consider the graph of a function $f(x)$, whose derivative $f'(x)$ and the second derivative $f''(x)$, see Fig. 3.7. Local extrema and turning points can be determined from the slope and curvature behavior.

Local Extrema

First, let's look at the point $x = x_M$. The function $f(x)$ has a **local maximum** at this point. The necessary condition for this is that $f'(x_M) = 0$. As can be seen from the graph of $f'(x)$, $f'(x)$ has a sign change from + to − in the case of a maximum. This is the sufficient condition for the existence of a maximum. For a **local minimum**, the corresponding conditions apply: The necessary condition is $f'(x) = 0$, the sufficient condition is a sign change from − to + in the derivative function $f'(x)$.

Note:

In many cases, a look at the second derivative $f''(x)$ is helpful. The requirement of the sufficient condition for a local extremum is $f''(x) \neq 0$. For $f''(x) < 0$, a local maximum results, and for $f''(x) > 0$, a local minimum results. However, this is not always the case, as the example $f(x) = x^6$ shows. Obviously, in $x_0 = 0$, there is a local minimum, but the second derivative $f''(x) = 30x^4$ has the value 0 at this point. Therefore, in case of doubt, it is advisable to look at the change in sign of the first derivative $f'(x)$.

In addition to the local extrema, **global extrema** are often sought. To find the **global minimum**, one compares the smallest local minimum with the boundary values of the interval on which the function is defined and chooses the minimum here. The procedure is analogous when searching for the **global maximum**.

Turning Points

Let us now turn to the point $x = x_W$ in Fig. 3.7.

3.5 Higher Derivatives, Extreme Values and Turning Points

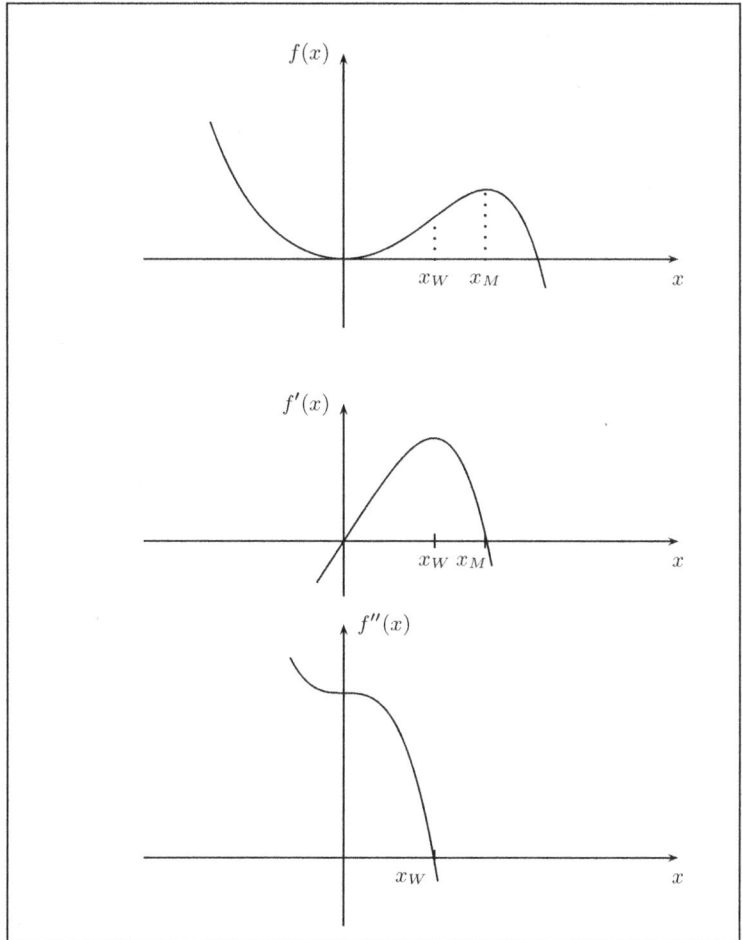

Fig. 3.7 To determine extrema and turning points

The slope $f'(x)$ has a local maximum at x_W. The graph of the function $f(x)$ has a **turning point** at this point $x = x_W$. The second derivative $f''(x)$ is obviously 0 at x_W. This is the necessary condition for a turning point with the sufficient condition of the change of sign of $f''(x)$. The second derivative $f''(x)$ characterizes the *curvature behavior* of $f(x)$. At the turning point, the transition from a positive curvature (or left curve) to a negative curvature (or right curve) takes place (or vice versa). Areas of positive curvature are also referred to as *convex from above*, areas with negative curvature are called *concave from above*.[4]

---

[4] See also Spivak (2008), Chap. 11.

To discuss global extreme values and turning points, we use the example from Sect. 2.2.12 for the production function.

Given the revenue-generating production function

$$x(r) = -r^3 + 7r^2 + 12r$$

with the domain $0 \leq r \leq 7$. We first form the derivatives

$$\frac{dx}{dr} = -3r^2 + 14r + 12$$

and

$$\frac{d^2x}{dr^2} = -6r + 14.$$

The necessary condition for the existence of local extrema is:

$$\frac{dx}{dr} \stackrel{!}{=} 0 \iff -3r^2 + 14r + 12 = 0$$

with the two solutions

$$r_{1,2} = \frac{-14 \pm \sqrt{196 + 144}}{-6} = \frac{-7 \pm \sqrt{85}}{-3},$$

thus:

$$r_1 \approx 5{,}4, \quad r_2 \approx -0{,}74, r_2 < 0.$$

Because of

$$\left.\frac{d^2x}{dr^2}\right|_{r=r_1} < 0$$

there is a local maximum at r_1 with

$$x(r_1) = 111{,}46.$$

To determine the global maximum, we now have to consider the boundary values of the domain:

$$x(0) = 0 \quad \text{and} \quad x(7) = 84.$$

Thus, the point $(5, 4|111, 46)$ is the global maximum of the function in the range $[0, 7]$.

To determine the turning point, we set:

$$\frac{d^2x}{dr^2} \stackrel{!}{=} 0 \iff -6r + 14 = 0 \text{ hence } r = \frac{7}{3}.$$

The second derivative $\frac{d^2x}{dr^2}$ has a sign change there, so there is actually a turning point there.

3.6 Applications of Differential Calculus

This section presents applications of differential calculus. These include Taylor series, limit determination with the rule of de L'Hospital, the Newton method for calculating zeros, curve discussion, and some economic considerations.

3.6.1 L'Hospital's Rule

The derivative function can be used to determine the limits of indeterminate expressions (e.g. of the form $\frac{0}{0}$).

L'Hospital's Rule:
Let $f(x)$, $g(x)$ be continuous functions with $f(x_0) = g(x_0) = 0$. Then:

$$\lim_{x \to x_0} \frac{f(x)}{g(x)} = \lim_{x \to x_0} \frac{f'(x)}{g'(x)} = \frac{f'(x_0)}{g'(x_0)}.$$

Proof

$$\lim_{x \to x_0} \frac{f(x)}{g(x)} = \lim_{\Delta x \to 0} \frac{f(x_0 + \Delta x)}{g(x_0 + \Delta x)}$$

because $f(x_0) = g(x_0) = 0$

$$= \lim_{\Delta x \to 0} \frac{f(x_0 + \Delta x) - f(x_0)}{g(x_0 + \Delta x) - g(x_0)}$$

$$= \lim_{\Delta x \to 0} \frac{f(x_0 + \Delta x) - f(x_0)}{g(x_0 + \Delta x) - g(x_0)} \cdot \frac{\Delta x}{\Delta x}$$

$$= \lim_{\Delta x \to 0} \frac{\frac{f(x_0 + \Delta x) - f(x_0)}{\Delta x}}{\frac{g(x_0 + \Delta x) - g(x_0)}{\Delta x}}$$

$$= \frac{f'(x_0)}{g'(x_0)}.$$

Examples

1. First, we verify the rule of de L'Hospital with an example where the limit can also be determined by elementary transformations. Let

$$u(x) = \frac{x^2}{2x},$$

be the behavior of this function in the limit $x \to 0$ has the form:

$$\lim_{x \to 0} u(x) = \lim_{x \to 0} \frac{x^2}{2x} = \frac{0}{0}.$$

By cancelling out common factors, we achieve:

$$\lim_{x \to 0} \frac{x}{2} = 0.$$

According to de L'Hospital we get:

$$\lim_{x \to 0} u(x) = \lim_{x \to 0} \frac{2x}{2} = 0.$$

2. In the following example, the limit can no longer be determined by a simple transformation of the expression. With the rule of de L'Hospital we get for:

$$\lim_{x \to 0} \frac{e^x - 1}{x} = \lim_{x \to 0} \frac{e^x}{1}$$
$$= e^0$$
$$= 1.$$

3. Let

$$f(x) = \frac{\ln x}{2(x-1)^2},$$

be considered here at the point $x_0 = 1$, then the indeterminate expression results:

$$\lim_{x \to 1} f(x) = \lim_{x \to 1} \frac{\ln x}{2(x-1)^2} = \frac{0}{0}.$$

The application of the rule of de L'Hospital gives:

$$\lim_{x \to 1} f(x) = \lim_{x \to 1} \frac{\frac{1}{x}}{4(x-1)}$$
$$= \lim_{x \to 1} \frac{1}{4x(x-1)}$$
$$= \infty.$$

The rule of de L'Hospital can be extended to other cases, in particular to limits for $x \to \pm\infty$ and indeterminate expressions of the form $\frac{\infty}{\infty}$ (see Erwe (1962) pp. 161 ff.).

Note:
The application of the rule of de L'Hospital is subject to certain conditions. If these conditions are not met, the application of the rule of de L'Hospital leads to false results, as the following example shows:
Let

$$f(x) = \frac{e^x - 2}{x},$$

then:

$$\lim_{x \to 0} \frac{e^x - 2}{x} = -\infty,$$

since there is a constant in the numerator and the denominator goes to zero. If we apply **inadmissibly** the rule of de L'Hospital here,

$$\lim_{x \to 0} \frac{e^x - 2}{x} = \frac{\lim_{x \to 0} e^x}{\lim_{x \to 0} 1} = 1,$$

we get a false result!

3.6.2 Determination of Zeros with the Newton Method

Another application of differential calculus arises in the determination of zeros of functions. A very frequently occurring problem is:

Given a function $y = f(x)$ with $x \in D_f$. The task is to find those $x_i \in D_f$ with $f(x_i) = 0$.

In Fig. 3.8 the graph of the function $f(x) = 0.5x^5 + 1.3x^4 + x^3 + x^2 - 0.1$ is shown. The zeros of this function are the x_i at which the graph intersects the x-axis.

Determining the zeros is no problem if the function is linear or quadratic. See, for example, Sects. 2.2.1 and 2.2.2.

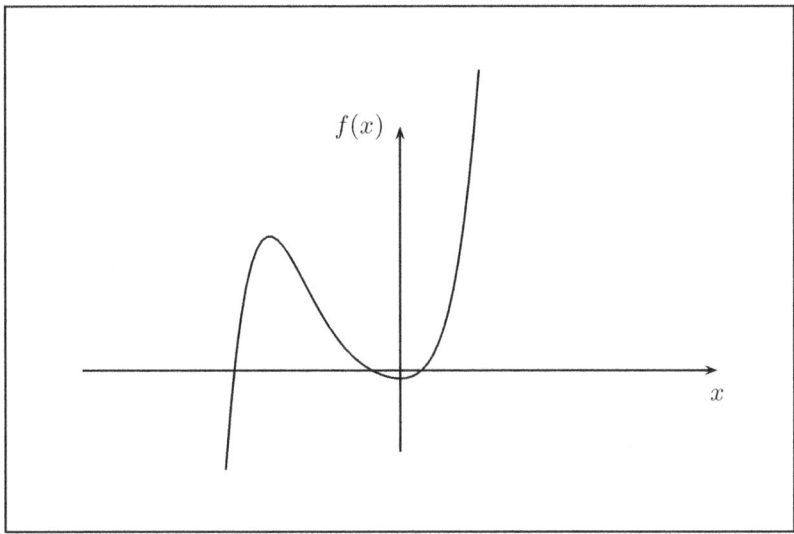

Fig. 3.8 The graph of the function $f(x) = 0.5x^5 + 1.3x^4 + x^3 + x^2 - 0.1$

For other functions, the calculation of zeros becomes difficult. For polynomials of higher order, the factorization is used. However, this is only helpful if at least one zero is easy to find.

Example Consider the cubic polynomial function:

$$f(x) = x^3 + x^2 - 6x + 4.$$

The x_i with $f(x_i) = 0$ are searched for. Therefore, the equation to be solved is:

$$x^3 + x^2 - 6x + 4 = 0.$$

By trial and error, one finds a first zero $x = 1$. With this, one sets

$$f(x) = (x - 1) \cdot g(x)$$

or

$$g(x) = \frac{f(x)}{x - 1}.$$

Now the polynomial $x^3 + x^2 - 6x + 4$ must be divided by the polynomial $x - 1$. It follows:

$$(x^3 + x^2 - 6x + 4) : (x - 1) = x^2 + 2 \cdot x - 4 = g(x),$$

because:

$$
\begin{array}{l}
(x^3 + x^2 - 6x + 4) : (x - 1) = x^2 + 2x - 4 \\
\underline{-(x^3 - x^2)} \\
\quad 2x^2 - 6x \\
\underline{\quad -(2x^2 - 2x)} \\
\quad\quad -4x + 4 \\
\underline{\quad\quad -(-4x + 4)} \\
\quad\quad\quad 0.
\end{array}
$$

This reduces the problem of determining zeros to determining the zeros of a parabola. The quadratic equation

$$x^2 + 2x - 4 = 0$$

leads to the solutions:

$$x_{1/2} = \frac{-2 \pm \sqrt{4 + 4 \cdot 4}}{2} = -1 \pm \sqrt{5}.$$

This method of finding zeros fails if a first zero cannot be found by trial and error or if equations such as:

3.6 Applications of Differential Calculus

$$x - \sin x = 0$$

are involved. Here, with the exception of the point $x_0 = 0$, no closed solution (i.e. a formula for the zeros) can be given. Therefore, in such cases, one has to resort to **approximation methods**. Approximation methods usually work **iteratively**. You start with an initial solution as an approximation for the sought zero and then improve this solution step by step. A method we will not go into here is the *regula falsi* (see Courant 1971a).

We want to look at the **Newton method**, which takes into account not only the function value at the approximate location, but also the slope. This is done as follows: x_1 is the 1st approximation for the sought-after zero point. The tangent through the point $P_1 = (x_1, f(x_1))$ is cut with the x-axis. This intersection provides the 2nd approximation for the sought-after zero point. Again, the tangent to f is cut at $P_2 = (x_2, f(x_2))$ with the x-axis and one obtains the next approximation for the zero point. This procedure can be repeated and one approaches the zero point **iteratively**. Therefore, the Newton method for determining zeros is an **iterative approximation method**.

The quantification of these considerations looks as follows (Fig. 3.9):

Given are a function $y = f(x)$ and an initial value x_1. Sought is the tangent at point P_1. We call this tangent t_1. Obviously t_1 is given by a general linear equation of the form:

$$t_1(x) = mx + b.$$

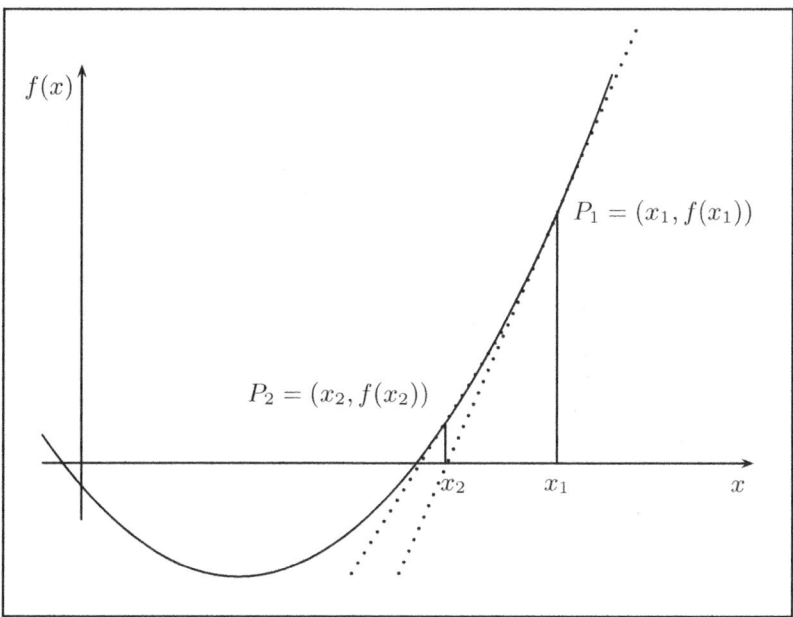

Fig. 3.9 For the construction of the Newton method

Now the two unknowns m and b must be determined. This is done through the following two conditions:

1. t_1 has the same slope as the function $f(x)$ at point x_1. This means:
$$m = f'(x_1).$$
It follows:
$$t_1(x) = f'(x_1)x + b.$$

2. The tangent $t_1(x)$ intersects the curve $f(x)$ at point x_1. This condition implies:
$$t_1(x_1) = f(x_1)$$
or
$$f(x_1) = f'(x_1)x_1 + b$$
or
$$b = f(x_1) - x_1 \cdot f'(x_1).$$

This leads to the equation of the line:
$$t_1(x) = f'(x_1)x + f(x_1) - x_1 f'(x_1).$$

We are now looking for the position x_2. This is the point at which the tangent t_1 intersects the x-axis:
$$t_1(x_2) = 0.$$

If we insert the above derived equation of the tangent into this condition, it follows:
$$f'(x_1)x_2 + f(x_1) - x_1 f'(x_1) = 0.$$

If you solve this equation for x_2, it follows (Fig. 3.9):
$$\boxed{x_2 = x_1 - \frac{f(x_1)}{f'(x_1)}.} \tag{3.16}$$

Now you replace x_1 by x_2, P_1 by P_2 and determine the next approximation completely analogously to:
$$\boxed{x_3 = x_2 - \frac{f(x_2)}{f'(x_2)}.}$$

In general:

3.6 Applications of Differential Calculus

The $k+1$-th approximation for a zero of the function $f(x)$ results iteratively from the k-th approximation

$$x_{k+1} = x_k - \frac{f(x_k)}{f'(x_k)}. \tag{3.17}$$

Examples First, we demonstrate the Newton method using an example in which we already know the zero. We consider the function:

$$f(x) = e^x - 1.$$

This function has a zero in $x = 0$.

Now we apply the Newton method and start in 1st approximation in $x_1 = 1$. Then it is:
The 2nd approach:

$$x_2 = x_1 - \frac{f(x_1)}{f'(x_1)}$$

$$= 1 - \frac{e^1 - 1}{e^1}$$

$$= 1 - 1 + \frac{1}{e}$$

$$= e^{-1}$$

$$\approx 0,3678.$$

The 3rd approach:

$$x_3 = x_2 - \frac{f(x_2)}{f'(x_2)}$$

$$= e^{-1} - \frac{e^{e^{-1}} - 1}{e^{e^{-1}}}$$

$$= e^{-1} - 1 + \frac{1}{e^{e^{-1}}} \approx 0,06.$$

As a second example—here the zero point is not immediately visible—we consider the function

$$f(x) = x - \cos(x).$$

In Fig. 3.10 the graphs of the two functions $h(x) = x$ and $g(x) = \cos x$ are shown in a coordinate system. This representation shows that these two functions have exactly one point of intersection at the point $x = \widetilde{x}$ Therefore, the function $f(x) = h(x) - g(x) = x - \cos x$ has exactly one zero point.

The derivative of this function is:

$$f'(x) = 1 + \sin(x)$$

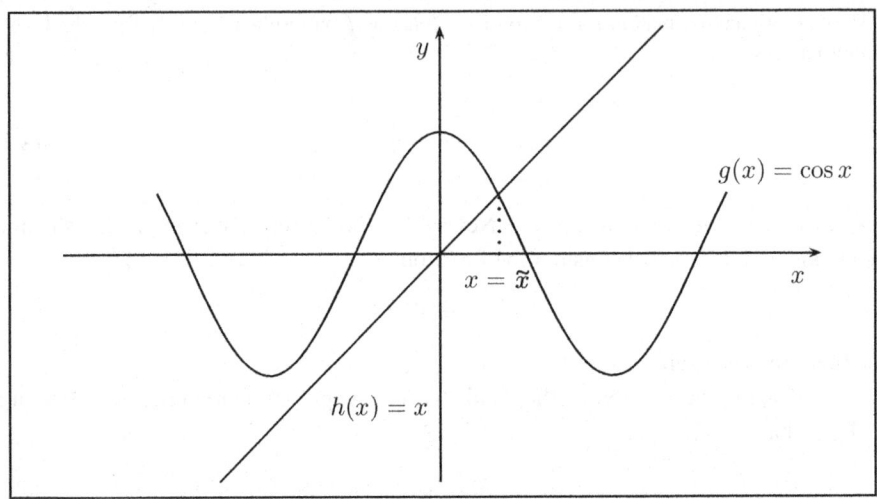

Fig. 3.10 Graphs of the functions $h(x) = x$ and $g(x) = \cos(x)$. Here it is clear that the two functions have exactly one point of intersection

The task is to find the zero point using the Newton method. Let's apply the Newton method and start in the 1st approximation in $x_1 = 1$. Then:

$$x_2 = x_1 - \frac{f(x_1)}{f'(x_1)}$$
$$= 1 - \frac{x - \cos(x)}{1 + \sin(x)} \bigg|_{x_1 = 1}$$
$$= 1 - \frac{1 - \cos(1)}{1 + \sin(1)}$$
$$\approx 0{,}75036.$$

The second approximation results with some calculator help to:

$$x_3 = x_2 - \frac{f(x_2)}{f'(x_2)}$$
$$= 0{,}75036 - \frac{0{,}75036 - \cos(0{,}75036)}{0{,}75036 + \sin(0{,}75036)}$$
$$\approx 0{,}7391.$$

The Newton method delivers depending on the start value x_1 (first approximation) only one zero of the function. To find more zeros, another start value must be chosen.

Example We consider the function:

$$f(x) = h(x) - g(x) = x - 3\cos\frac{3x}{2}.$$

3.6 Applications of Differential Calculus

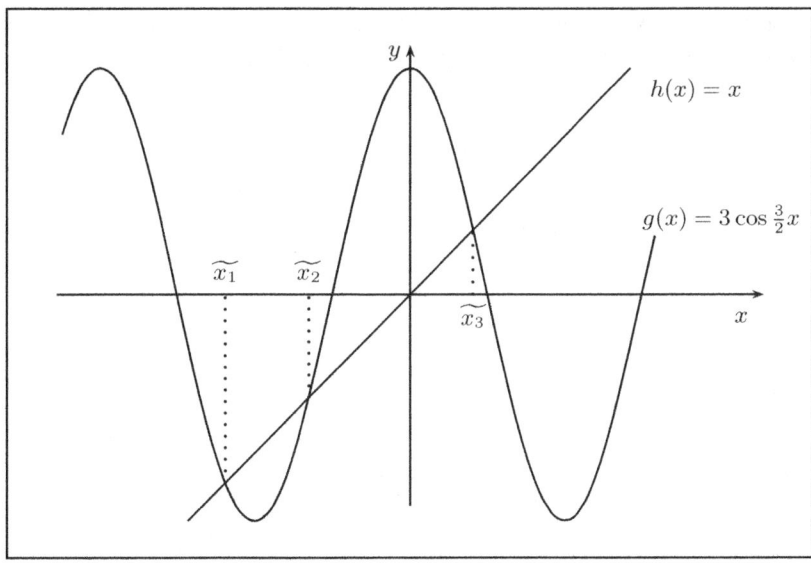

Fig. 3.11 Graphs of the functions $h(x) = x$ and $g(x) = 3\cos 3x/2$ with three intersection points $\widetilde{x_1} \approx -2{,}489, \widetilde{x_2} \approx -1{,}36$ and $\widetilde{x_3} \approx 0{,}854$

As Fig. 3.11 shows, these two functions have three interfaces, so the function $f(x) = h(x) - g(x)$ has three zeros $\widetilde{x_1}, \widetilde{x_2}$ and $\widetilde{x_3}$. In this example, the approximation for one of the three zeros is very sensitive to the initial value. For example, if the initial value is $x_0 = 3$, the Newton method leads to the solution $\widetilde{x_2} \approx -1{,}36$. This example shows that the Newton method does not necessarily provide the solution closest to the initial value.

Termination Criterion for the Newton Method

As with any approximation method, the Newton method must be terminated in an appropriate form. The following options are available:

- A number of iterations is specified
- $|f(x)|$ must be less than a given threshold
- $|x_{k+1} - x_k|$ becomes less than a threshold $\delta > 0$.

3.6.3 Taylor Series

We are looking for a way to approximate any elementary function such as sin, cos, ln or exp by a polynomial

$$P(x) = a_0 + a_1 x + a_2 x^2 + \cdots + a_n x^n, \tag{3.18}$$

with real coefficients $a_i, i = 1, 2, \ldots, n$.

We consider a polynomial function of the form (3.18). As is easily seen, the coefficients a_i can be expressed by the value of P and the higher derivatives at the point $x = 0$. So we get

$$P(0) = a_0, \quad \text{or} \quad a_0 = \frac{P^{(0)}(0)}{0!}. \tag{3.19}$$

If we differentiate Eq. (3.18) with respect to x, we get

$$P'(x) = a_1 + 2a_2 x + \cdots + n \cdot a_n x^{n-1},$$

therefore[5]

$$P'(0) = P^{(1)}(0) = a_1 \quad \text{or} \quad a_1 = \frac{P^{(1)}(0)}{1!}.$$

This can be generalized, for the coefficient a_k we get:

$$P^{(k)}(0) = k! \, a_k, \quad \text{or} \quad a_k = \frac{P^{(k)}(0)}{k!}. \tag{3.20}$$

The expression (3.19) shows that (3.20) is true for $k = 0, 1, \ldots, n$. If we start with a polynomial of the form

$$P(x) = a_0 + a_1(x - x_0) + a_2(x - x_0)^2 + \cdots + a_n(x - x_0)^n, \tag{3.21a}$$

then this argumentation leads to the coefficients

$$a_k = \frac{P^{(k)}(x_0)}{k!}. \tag{3.21b}$$

Let f be a function.[6] The n derivatives of the function $f(x)$

$$f^{(1)}(x_0), f^{(2)}(x_0), \ldots, f^{(n)}(x_0)$$

at the point $x_0 \in D_f$ exist. If we define coefficients

$$a_k = \frac{f^{(k)}(x_0)}{k!}, \quad \text{where} \quad 0 \le k \le n,$$

then we can consider the following polynomial function of order n

$$P_n(x) = a_0 + a_1(x - x_0) + a_2(x - x_0)^2 + \cdots + a_n(x - x_0)^n$$

$$= f(x_0) + \frac{f^{(1)}(x_0)}{1!}(x - x_0) + \frac{f^{(2)}(x_0)}{2!}(x - x_0)^2$$

$$+ \cdots + \frac{f^{(n)}(x_0)}{n!}(x - x_0)^n.$$

---

[5] Note that the notation P(k)(x) denotes the k-th derivative of the polynomial P(x); do not confuse it with the power of a function.

[6] Not necessarily a polynomial.

3.6 Applications of Differential Calculus

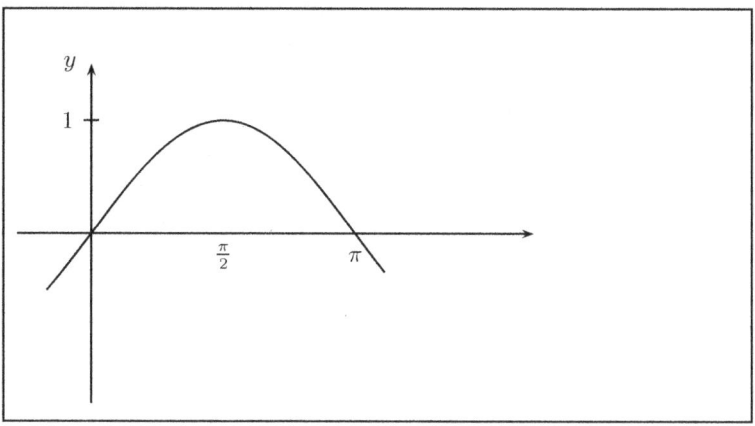

Fig. 3.12 The function $f(x) = \sin x$

Example We consider the function $f(x) = \sin x$ at the point $x_0 = \pi/2$ (Fig. 3.12).

The graph of this function around the point $x_0 = \pi/2$ suggests that the function can be approximated in the vicinity of this point by a—downward open—parabola.

We therefore make the ansatz for the polynomial:

$$P(x) = a_0 + a_1\left(x - \frac{\pi}{2}\right) + a_2\left(x - \frac{\pi}{2}\right)^2$$
$$+ a_3\left(x - \frac{\pi}{2}\right)^3 + a_4\left(x - \frac{\pi}{2}\right)^4,$$

with five parameters $a_0, \ldots, a_4$. We determine these parameters from the conditions

$f\left(\frac{\pi}{2}\right) = 1$ and $P\left(\frac{\pi}{2}\right) = a_0$ $\implies$ $a_0 = 1$.

$\frac{df(\pi/2)}{dx} = 0$ and $\frac{dP(\pi/2)}{dx} = a_1$ $\implies$ $a_1 = 0$.

$\frac{d^2 f(\pi/2)}{dx^2} = -1$ and $\frac{d^2 P(\pi/2)}{dx^2} = 2a_2$ $\implies$ $a_2 = -\frac{1}{2}$.

$\frac{d^3 f(\pi/2)}{dx^3} = 0$ and $\frac{d^3 P(\pi/2)}{dx^3} = 6a_3$ $\implies$ $a_3 = 0$.

$\frac{d^4 f(\pi/2)}{dx^4} = 1$ and $\frac{d^4 P(\pi/2)}{dx^4} = 24a_4$ $\implies$ $a_4 = \frac{1}{24}$.

This gives the desired polynomial (up to 4th order):

$$P_4(x) = 1 - \frac{1}{2}\left(x - \frac{\pi}{2}\right)^2 + \frac{1}{24}\left(x - \frac{\pi}{2}\right)^4. \tag{3.22}$$

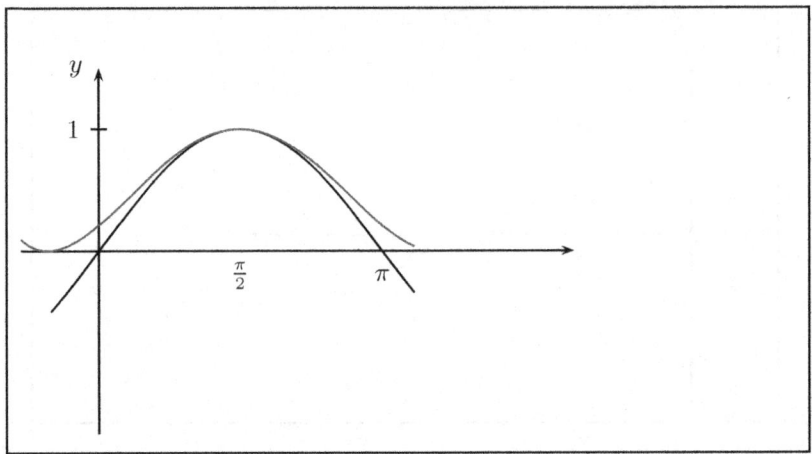

Fig. 3.13 Approximation of the sine function by a polynomial

Figure 3.13 shows the function $f(x)$ and the graph of the polynomial (3.22) in a diagram. The polynomial $P_4(x)$ is a very good approximation of the function $f(x)$ in a neighborhood of the point $x_0 = \pi/2$.

▶ **Definition (Taylorpolynom)**

Let

$$f : D \subseteq \mathbb{R} \longrightarrow \mathbb{R}$$

n times differentiable on an open interval $D \subseteq \mathbb{R}$. Then

$$P_n(x) = \sum_{k=0}^{n} \frac{f^{(k)}(x_0)}{k!}(x-x_0)^k, \quad x \in \mathbb{R} \qquad (3.23)$$

the **Taylorpolynom** denotes the Taylor polynomial of degree n of the function f at x_0.[7]
Example $f(x) = \cos x$ at $x_0 = 0$. The derivatives of the cosine function are:

$$
\begin{aligned}
k=0: & \quad f^{(0)}(x) = \cos x & \Longrightarrow f^{(0)}(0) = 1, \\
k=1: & \quad f^{(1)}(x) = -\sin x & \Longrightarrow f^{(1)}(0) = 0, \\
k=2: & \quad f^{(2)}(x) = -\cos x & \Longrightarrow f^{(2)}(0) = -1, \\
k=3: & \quad f^{(3)}(x) = \sin x & \Longrightarrow f^{(3)}(0) = 0, \\
k=4: & \quad f^{(4)}(x) = \cos x & \Longrightarrow f^{(4)}(0) = 1.
\end{aligned}
$$

---

[7] This polynomial is named after Brook Taylor (1685–1731), who published this result in 1715. See Katz (2009) or Merzbach and Boyer(2011).

3.6 Applications of Differential Calculus

From here on, the values of the derivatives repeat with a period of 4. Therefore, the coefficients of the Taylor polynomial are:

$$1, 0, -\frac{1}{2!}, 0, \frac{1}{4!}, 0, -\frac{1}{6!}, 0, \frac{1}{8!}, \ldots$$

This gives the Taylor polynomial:

$$P_{2n}(x) = 1 - \frac{x^2}{2!} + \frac{x^4}{4!} - \frac{x^6}{6!} + \cdots + (-1)^n \frac{x^{2n}}{(2n)!} = \sum_{k=0}^{n} (-1)^k \frac{x^{2k}}{(2k)!}. \quad (3.24)$$

The finite series (3.24) is the Taylor polynomial of degree $2n$ of the function $f(x) = \cos x$ at the point $x_0 = 0$.

Example The Taylor polynomial of the exponential function e^x is particularly simple. Since

$$\left. \frac{d^k}{dx^k} e^x \right|_{x=0} = 1 \qquad \text{for all } k = 0, 1, 2, 3, \ldots.$$

This means that the Taylor polynomial of degree n of the exponential function for the development point $x_0 = 0$:

$$P_n(x) = 1 + \frac{x}{1!} + \frac{x^2}{2!} + \frac{x^3}{3!} + \cdots + \frac{x^n}{n!} = \sum_{k=0}^{n} \frac{x^k}{k!}. \quad (3.25)$$

The question now is how good the approximation of a function by its Taylor polynomial of degree n is. For this we write:

$$f(x) = P_n(x) + R_{n+1}(x, x_0). \quad (3.26)$$

The term $R_{n+1}(x, x_0)$ in Eq. (3.26) is called the **remainder**; this term describes the difference between the function and the Taylor polynomial. The remainder depends on the location x_0, the order of the polynomial n and the argument x. Eq. (3.26) is called the **Taylor formula**.

The remainder is

$$R_{n+1}(x, x_0) = \frac{f^{(n+1)}(c)}{(n+1)!} (x - x_0)^{n+1}, \quad (3.27)$$

where c is a point between x and x_0.[8] The form of the remainder in Eq. (3.27) is called **Lagrange remainder representation**.

---

[8] For a proof and further aspects, we refer to the literature, e.g. Arens et al. (2018), Chapter 10, Lang (1986), Chapter XIII or Spivak (2008), Chapter 20. Very detailed investigations can be found in Marsden and Weinstein (1985), Calculus II, Chapter 12.

This gives us the Taylor formula to:

$$f(x) = f(x_0) + \frac{f^{(1)}(x_0)}{1!}(x - x_0) + \frac{f^{(2)}(x_0)}{2!}(x - x_0)^2 \\ + \cdots + \frac{f^{(n)}(x_0)}{n!}(x - x_0)^n + R_{n+1}(x, x_0). \quad (3.28)$$

Here we assume that f is a n times continuously differentiable function and $x, x_0 \in D_f$.

If the function $f(x)$ is differentiable any number of times, the index n in the Taylor formula (3.28) can be run to infinity. Provided that the function $f(x)$ is differentiable any number of times and the remainder $R_{n+1}(x, x_0)$ goes to 0 for $n \to \infty$, we obtain the **Taylor series** of the function $f(x)$.

Taylor and Maclaurin Series

If f is a function that is differentiable any number of times on an interval D, which contains the point x_0, then the series

$$f(x) = \sum_{n=0}^{\infty} \frac{f^{(n)}(x_0)}{n!}(x - x_0)^n \quad (3.29)$$

is called the Taylor series of f at the point x_0.[9] The series (3.29) is also called the **power series expansion** of the function $f(x)$. If $x_0 = 0$, then the series has the simpler form

$$f(x) = \sum_{n=0}^{\infty} \frac{f^{(n)}(0)}{n!} x^n, \quad (3.30)$$

the series (3.30) is called the Maclaurin series of f.[10]

Examples The Taylor series of the sine function is:

$$\sin x = \sum_{n=0}^{\infty} \frac{(-1)^n}{(2n+1)!} x^{2n+1}. \quad (3.31a)$$

The Taylor series of the cosine function is

$$\cos x = \sum_{n=0}^{\infty} \frac{(-1)^n}{(2n)!} x^{2n}. \quad (3.31b)$$

---

[9] This series is named after the British mathematician Brook Taylor (1685–1731), who published it for the first time in 1715.

[10] This series is named after the British mathematician Colin Maclaurin (1698–1746).

3.6 Applications of Differential Calculus

The Taylor series of the e function is

$$e^x = \sum_{n=0}^{\infty} \frac{1}{n!} x^n. \tag{3.31c}$$

The power series (3.31a) and (3.31b) show an important property of Taylor series. The function $\sin x$ is odd, i. e. $\sin(-x) = -\sin x$. The Taylor series of the sine function shows the same property, because the series contains only odd powers of x. Analogously, the cosine function is even, the Taylor series (3.31b) contains only even powers of x.

From the development of a function into its Taylor series, simple approximation functions of a function $f(x)$ in the form of polynomials can be obtained by truncating the series. This makes it possible, inter alia:

- the approximation of a function by a polynomial function,
- the approximate calculation of function values,
- the integration of a function (numerical integration), in Sect. 4.3.4 an example can be found,
- the calculation of limits of indeterminate expressions.

Example Let's look at the function

$$f(x) = \frac{x - \sin x}{x \cdot \sin x}. \tag{3.32}$$

We are looking for the behavior of this function for $x \to 0$. In addition to the rule of de L'Hospital, the Taylor series also allows the calculation of indeterminate expressions. The sine function has the Taylor expansion

$$\sin x = \sum_{n=0}^{\infty} \frac{(-1)^n}{(2n+1)!} \cdot x^{2n+1}$$
$$= x - \frac{x^3}{3!} + \frac{x^5}{5!} \pm \cdots. \tag{3.33}$$

If we insert this expansion in the numerator and denominator of the function (3.32), we get:

$$x - \sin x = x - x + \frac{x^3}{3!} - \frac{x^5}{5!} \pm \cdots = \frac{x^3}{3!} - \frac{x^5}{5!} \pm \cdots$$

and

$$x \cdot \sin x = x^2 - \frac{x^4}{3!} + \frac{x^6}{5!} \pm \cdots$$

Therefore:

$$\lim_{x \to 0} f(x) = \lim_{x \to 0} \frac{x - \sin x}{x \cdot \sin x}$$

$$= \lim_{x \to 0} \frac{\frac{x^3}{3!} - \frac{x^5}{5!} \pm \cdots}{x^2 - \frac{x^4}{3!} + \frac{x^6}{5!} \pm \cdots}$$

$$= 0.$$

3.6.4 Curve Discussion

Curve discussion deals with the question of how the graph of a function $f(x)$ looks. The following properties of a function are examined:

- Definition and range
- Symmetry
- Zero points
- Pole points
- Behavior for $|x| \to \infty$ (this is also called *asymptotic*) .

In addition, the application of differential calculus delivers:

- local and global extreme values
- turning points.

Examples
1. We first consider the rational function as an example of a curve discussion:

$$f(x) = \frac{x^2 - x - 2}{2x - 6}. \tag{3.34}$$

This function has the form:

$$f(x) = \frac{u(x)}{v(x)}$$

with

$$u(x) = x^2 - x - 2$$
$$v(x) = 2x - 6.$$

(a) Domain of definition:
The function is defined in all points where the denominator $v(x)$ is not equal to zero, therefore:

$$D_f = \mathbb{R} \setminus \{3\}.$$

3.6 Applications of Differential Calculus

(b) Symmetry:

$f(x)$ is neither point-symmetric nor axis-symmetric, since

$$f(-x) \neq f(x) \text{ and } f(-x) \neq -f(x).$$

(c) Determination of zeros:

The function $f(x)$ is 0 if $u(x)=0$. It holds:

$$u(x) = 0 \iff x^2 - x - 2 = 0$$
$$\iff (x-2)(x+1) = 0,$$

from which it follows that $x_1 = 2$ and $x_2 = -1$ are zeros of $f(x)$.

(d) Determination of poles:

The function $f(x)$ has poles exactly when $v(x)=0$. Since

$$v(x) = 2(x-3),$$

this is obviously the case for $x=3$.

The behavior of $f(x)$ at this pole can be analyzed using the following limit consideration:

$$\lim_{x \to 3+0} f(x) = \lim_{x \to 3+0} \frac{x^2 - x - 2}{2(x-3)}$$
$$= 2 \lim_{x \to 3+0} \frac{1}{x-3}$$
$$\to +\infty.$$

$$\lim_{x \to 3-0} f(x) = \lim_{x \to 3-0} \frac{x^2 - x - 2}{2(x-3)}$$
$$= 2 \lim_{x \to 3-0} \frac{1}{x-3}$$
$$\to -\infty.$$

Therefore there is a pole with a change of sign.

(e) Asymptotics for $x \to \pm\infty$:

To obtain the asymptotics of the function $f(x)$, we write $f(x)$ in the form:

$$f(x) = \frac{x - 1 - \frac{2}{x}}{2 - \frac{6}{x}}.$$

Then

$$\lim_{x \to \infty} f(x) = \frac{1}{2}(x-1)$$

with

$$\lim_{x \to +\infty} f(x) = +\infty$$
$$\lim_{x \to -\infty} f(x) = -\infty.$$

(f) Local extrema:
The necessary condition for the existence of a local extremum is $f'(x) = 0$. Using the quotient rule, one obtains:

$$f'(x) = \frac{(2x-1)(2x-6) - (x^2 - x - 2)2}{(2x-6)^2}$$
$$= \frac{2(2x-1)(x-3) - 2(x-2)(x+1)}{[2(x-3)]^2}.$$

Since $f'(x) = 0$ only when the numerator is zero, it follows that

$$(2x-1)(2x-6) - (x^2 - x - 2)2 = 0$$

or

$$x^2 - 6x + 5 = 0.$$

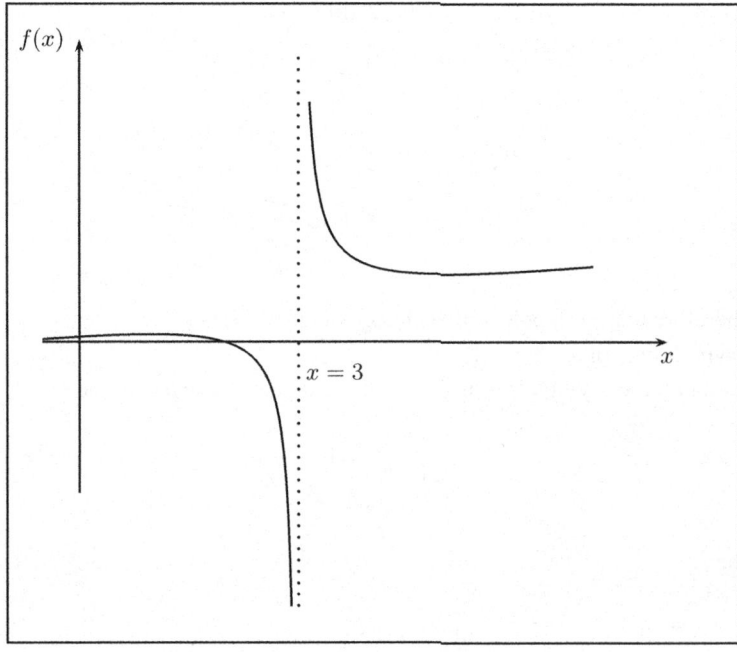

Fig. 3.14 The graph of the function (3.34)

3.6 Applications of Differential Calculus

This leads to the two solutions $x_1 = 1$ and $x_2 = 5$. These two x-values are candidates for extreme values. With the help of the second derivative, it can be checked whether they are extreme values or turning points. We spare ourselves this at this point, because together with the considerations on turning points and asymptotics it can be concluded that x_1 must be a high point and x_2 a low point. The graph of the function being examined (3.34) is shown in Fig. 3.14.

2. The calculation of extreme values plays an important role in economics. For obvious reasons, for example, the maximum of the revenue function is interesting.
   A revenue function is (see Fig. (2.5)):

$$E(p) = -cp^2 + x_0 p; \quad c > 0, x_0 > 0.$$

The derivative of this revenue function with respect to the price p yields:

$$\frac{dE(p)}{dp} = -2cp + x_0.$$

The revenue function is therefore extreme if:

$$\frac{dE(p)}{dp} = 0 \quad \Longleftrightarrow \quad p_{opt} = \frac{x_0}{2c}.$$

The second derivative of the revenue function results in:

$$\frac{d^2 E(p)}{dp^2} = -2c < 0 \quad \text{for all} \quad p.$$

This is the sufficient criterion for p_{opt} to be a local maximum. The revenue function itself has the value p_{opt} at the point:

$$E(p_{opt}) = \frac{1}{4} \frac{x_0^2}{c}.$$

3. The function

$$f(x) = \frac{1}{b\sqrt{2\pi}} e^{-\frac{(x-a)^2}{2b^2}}, \quad a, b \in \mathbb{R}, b > 0$$

is the Gaussian distribution, which plays a prominent role in probability theory and statistics (see Fig. 3.15).
The first derivative of this function is:

$$\frac{df(x)}{dx} = \left(-\frac{x-a}{b^2}\right) \frac{1}{b\sqrt{2\pi}} e^{-\frac{(x-a)^2}{2b^2}} \tag{3.35}$$

$$= \left(-\frac{x-a}{b^2}\right) f(x). \tag{3.36}$$

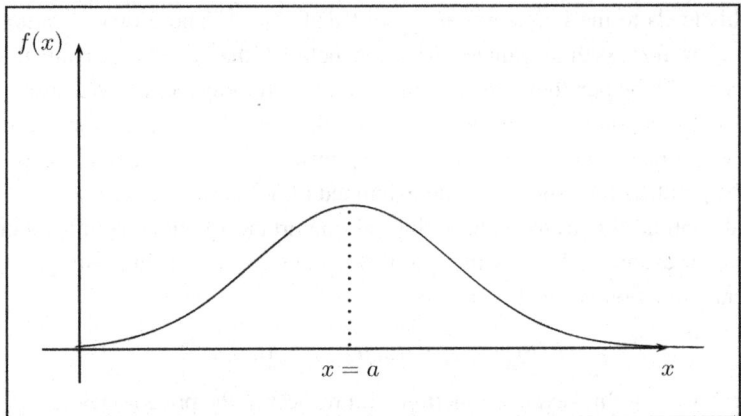

Fig. 3.15 The Gaussian distribution

Since $f(x) \neq 0$ for all $x \in \mathbb{R}$, we obtain the necessary condition for an extreme value as

$$f'(x) = 0 \iff x = a.$$

Therefore:

$$f(x = a) = \frac{1}{b\sqrt{2\pi}} e^{-\frac{(a-a)^2}{2b^2}}$$
$$= \frac{1}{b\sqrt{2\pi}} e^0$$
$$= \frac{1}{b\sqrt{2\pi}}.$$

The second derivative of the Gaussian distribution is:

$$\begin{aligned} f''(x) &= -\frac{f(x)}{b^2} - \left(\frac{x-a}{b^2}\right) f'(x) \\ &\stackrel{(3.36)}{=} -\frac{f(x)}{b^2} + \left(\frac{x-a}{b^2}\right)^2 f(x) \\ &= b^{-4} f(x) \left(x^2 - 2ax + a - b^2\right). \end{aligned}$$

Now

$$f''(a) = -\frac{f(a)}{b^2} = -\frac{1}{b^3 \cdot \sqrt{2\pi}} < 0,$$

therefore at the point $x = a$ there is a maximum. The turning points of this function are given by the zeros of the second derivative:

$$\frac{d^2 f(x)}{dx^2} = 0 \iff x^2 - 2ax + a^2 - b^2 = 0.$$

3.6 Applications of Differential Calculus

This is only the case if

$$x = a \pm b.$$

The size b is called *standard deviation* and is a measure of the width of the Gaussian distribution.

3.6.5 Limit Functions

The 1st derivative of a function characterizes the change in behavior of the function at a certain point x_0. If the function describes an economic relationship, then the term **limit function** is used for the 1st derivative.

▶ **Definition (Limit Function)**

The limit Function of an economic function f is the first derivative of this function.

The limit function thus describes the increase or decrease of the economic function.

First, we consider **marginal costs**. Starting from a cost function $K(x)$, which describes the production costs in dependence on the produced amount x (output), the marginal costs result to $K'(x)$. They play an important role in the consideration of the **profit function**. The profit function at a market-given price p (polypol) is

$$G(x) = p \cdot x - K(x).$$

The **profit zone** is given by the solutions of the inequality

$$p \cdot x > K(x).$$

The **maximum achievable profit** is obtained with the necessary condition $G'(x) = 0$ from the equation

$$p = K'(x).$$

Therefore, the marginal costs must be equal to the market price p in order to maximize profit.

The interesting question is the one about the **minimum price** p_{min}, which must be present on the market so that a provider makes a profit at all. Fig. 3.16 shows that the line $y = p_{min} \cdot x$ is the tangent of the cost function $K(x)$ at an as yet to be determined point x_B. The point of contact x_B results from the consideration

$$K'(x_B) = \frac{K(x_B) - 0}{x_B - 0} = \frac{K(x_B)}{x_B}, \qquad (3.37)$$

that is to say, the slope of the cost function at x_B is equal to the slope of the tangent. After the determination of x_B results in p_{min} to

$$p_{min} = K'(x_B).$$

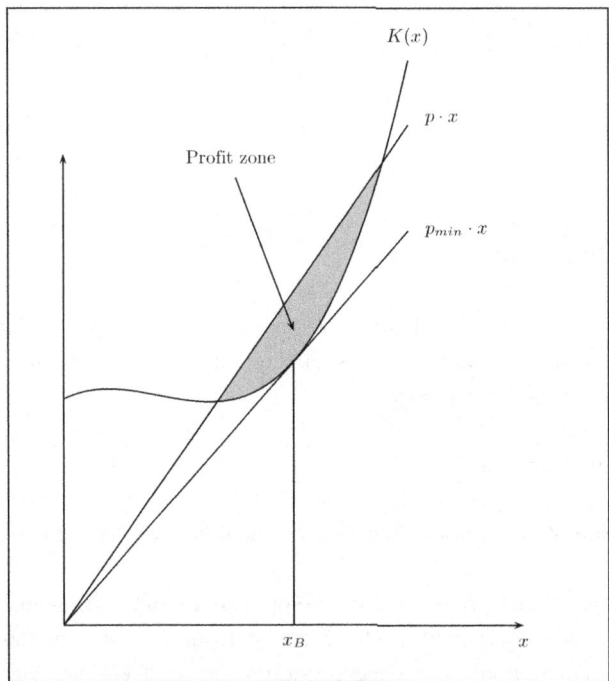

Fig. 3.16 Cost function $K(x)$ and revenue $p \cdot x$ in dependence on the produced amount x

The equation (3.37) can also be derived from the minimization of the unit costs (see Exercise 3.21).

More Limit Functions:
For the **revenue function** $E(x)$ considered in Sect. 2.2.12, the **marginal revenue** $\frac{dE}{dx}$ can be determined. Correspondingly, the **marginal productivity** results for the **production functions** considered in Sect. 2.2.12.

3.6.6 Elasticity of Functions

The derivative $f'(x)$ describes the absolute change of the function $f(x)$ at a change of the independent variable x at the point x_0. In economics, this is not always the decisive size, which has to be examined, in order to analyze a change behavior.

Frequently one is interested in **relative changes**. For example, one asks: By what percentage does the demand for a product change, if the price changes percentage-wise. The concept of **elasticity** is introduced to examine such questions .

3.6 Applications of Differential Calculus

▶ **Definition (Elasticity)**

The elasticity of the function $f(x)$ with respect to x is:

$$\epsilon_{f,x}(x) = \frac{\frac{df}{f}}{\frac{dx}{x}} \quad \text{where } x \neq 0, f \neq 0.$$

Because of:

$$df = f'(x)dx$$

we can write

$$\epsilon_{f,x}(x) = \frac{f'(x)}{f(x)} \cdot x.$$

The elasticity $\epsilon_{f,x}(x)$ approximately indicates by what percentage the function f changes, if x changes at the point x_0 by one percent. The approximation

$$\frac{df}{f} \approx \epsilon_{f,x}(x) \cdot \frac{dx}{x}$$

is valid for small relative changes dx/x.

Example For the function $f(x) = 4e^{-2x}$ the elasticity is given by:

$$\epsilon_{f,x}(x) = \frac{-8e^{-2x}}{4e^{-2x}} \cdot x = -2x.$$

Relative changes can be approximately described in dependence of the position x. If in this example for $x_0 = 2$ a change of 1% for x is carried out, then $\epsilon_{f,x}(2) = -4$, that means a decrease of 4%.

The following terms are introduced for different elasticity values:

Term	Notion	Meaning
$\lvert \epsilon_{f,x} \rvert > 1$	f is elastic	f changes more relative to x
$\lvert \epsilon_{f,x} \rvert < 1$	f is inelastic	f changes weaker relative to x
$\lvert \epsilon_{f,x} \rvert = 1$	f is proportionally elastic	f varies proportionally in the same way as x
Limit $\lvert \epsilon_{f,x} \rvert \to \infty$	f is totally elastic	Small changes in x lead to extreme changes in f
Limit $\lvert \epsilon_{f,x} \rvert = 0$	f is totally inelastic	f does not react to changes in x.

3.7 Exercises

Short solutions to the following exercises can be found in the appendix.

3.1 Determine the following limits:

(a) $\lim_{x\to 1} \frac{e^x-1}{\sqrt{x-1}}$.

(b) $\lim_{x\to 1} \frac{\ln x}{x-1}$.

3.2 Differentiate the following functions by applying the differentiation rules and check the result using the limit process:

(a) $y = -3x + 8$.
(b) $y = x^2 + a^2$.
(c) $y = (ax + b)^2$.

3.3 Determine the derivatives of the following functions:

(a) $f(x) = \frac{e^x}{x}$.
(b) $f(x) = \ln(2x^2 + 3x + 5)$.
(c) $f(x) = (\sqrt{x^2})^{1/5}$.
(d) $f(x) = x \ln x; \; x > 0$.

3.4 Show:

$$f'(x) = \frac{1}{x} \quad \text{for} \quad f(x) = \ln x, x > 0.$$

3.5 Examine $f(x)$ for continuity and differentiability at the interface:

(a) $f(x) = \begin{cases} x^2 + x & \text{if } x > 1 \\ 3x - 1 & \text{if } x \le 1. \end{cases}$

(b) $f(x) = \begin{cases} 1 + \ln x & \text{if } x \ge 1 \\ x^2 & \text{if } x < 1. \end{cases}$

3.6 Determine the intersection of $f(x) = \exp(-x)$ and $g(x) = x$ with the Newton method. Choose a suitable starting point and carry out one iteration step.

3.7 Determine the equations of the tangents to the graph of the function $f(x)$, which go through the point $P(-7/0)$ with

$$f(x) = \frac{2x - 1}{3x + 1}, \quad x \ne -\frac{1}{3}.$$

3.7 Exercises

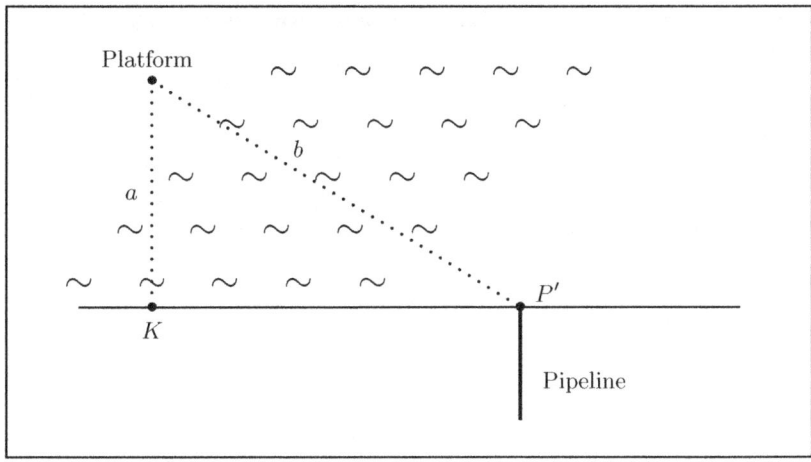

Fig. 3.17 Illustration of the situation in Excercise 3.8

3.8 An offshore oil platform is to be connected to the pipeline network on land (see Fig. 3.17). The platform is located at the distance $\overline{PK} = a$ from the coast and the distance $\overline{PP'} = b$ from the pipeline. Determine the cost-optimal connection of the platform to the pipeline if the laying costs on land *are* K_L and in the water K_W with ($K_W > K_L$).

3.9 Calculate the 2nd derivative of the following functions:

(a) $f(x) = x^3 - \frac{x^2}{2} + x - 1$.
(b) $f(x) = \exp(-x^2)$.
(c) $f(x) = 2\sin x + 5\cos x(2x+1)$.

3.10 Determine the coefficients of the general polynomial of 2nd degree so that $f(1)=3$ and $f'(0) = f''(0) = 1$.

3.11 What relationship must exist between the coefficients of

$$f(x) = ax^3 + bx^2 + cx + d$$

so that the graph of the function has no horizontal tangent?

3.12 Discuss the behavior of the function

$$f(x) = -x^3 + 2t^2x^2 + tx, x \in \mathbb{R}, t \in \mathbb{R}_0^+$$

in terms of zeros, extrema, behavior for large x, symmetry, and concave and convex regions.

3.13 For which $a, b \in \mathbb{R}$ is the function $f(x)$ continuously differentiable?

$$f(x) = \begin{cases} ax^2 - 2x & \text{if } x \geq 2 \\ bx + 1 & \text{if } x < 2. \end{cases}$$

3.14 Construct a function of 4th degree whose graph is symmetrical to the y-axis. A turning point has the coordinates $P = (1, 0)$. The two turning tangents intersect at right angles. Determine all zeros and extreme values of this function.

3.15 Calculate the Maclaurin series of the function

$$f(x) = \frac{1}{1-x}.$$

3.16 Calculate the Taylor series of the logarithm function ln at the point $x_0 = 1$.

3.17 Calculate the Taylor series of the function

$$f(x) = e^x \cdot \sin x$$

at the point $x = 0$.

(a) Develop the function $f(x)$ by forming the higher derivatives.
(b) Use the known power series of the exponential and the sine function.

3.18 Given is the function

$$f(x) = \cos x + e^{-x} - x^2.$$

Determine the zeros of this function using the Taylor polynomial $P_2(x)$.

3.19 Given is the price-demand function

$$x(p) = 10e^{-0.1p}.$$

(a) Determine the price elasticity.
(b) What is the change in demand when the price is lowered from 500 € to 495 € or when the price is raised from 500 € to 510 €?
(c) At what price does an increase of one percent lead to a decrease in demand of 10%?

3.20 Give examples of products with an elastic and an inelastic price elasticity.

3.21 Show: The minimum of the unit cost results from the intersection of the marginal cost and the unit cost. Illustrate this situation using the example

$$K(x) = 10 + 2x^2.$$

3.22 Determine the fixed cost component of a cost function if the minimum of the unit cost is assumed to be $x_0 = 25$ and the variable unit costs can be described by the function

$$k_V(x) = 10\sqrt{x} + 16$$

3.23 Determine the revenue-price function $E = E(p)$, in which the maximum revenue of 100,000 € is achieved for a price of $p = 100$ €. For the price-sales function, assume a linear relationship.

3.24 Given the cost function

$$K(x) = x^3 - 5x^2 + 6x + 48.$$

(a) Calculate the profit zone at a constant market price of $p = 24$ €.
(b) Determine the maximum possible profit.
(c) How high must the market price be at least so that a provider can make a profit at all?

3.25 Determine the parameters a and k so that the following function $f(x)$ is continuous and differentiable:

$$f(x) = \begin{cases} e^{k \cdot x} & \text{if } x \leq 1, \\ a \cdot \sqrt{x} & \text{if } x > 1. \end{cases}$$

3.26 A power supplier has a tariff model in which there is a basic price p_g and a price p_v that depends on consumption x. The revenue per customer is:

$$E(x) = p_g + p_v \cdot x, \quad (x > 0).$$

The cost function is given by

$$K(x) = \begin{cases} 6 \cdot \ln(x) + 3 & \text{if } x > 1, \\ 3 & \text{if } 0 \leq x \leq 1. \end{cases}$$

How high must p_v be chosen so that no loss is made at any consumption value x ? The basic price p_g is 4 units.

References

Alten H.-W., Djafari Naini A., Eick B., Folkerts M., Schlosser H., Schlote K.-H., Wesemüller-Kock H., Wußing H. (2014): 4000 Jahre Algebra, Geschichte Kulturen Menschen, 2. Auflage, Springer Verlag, Berlin, Heidelberg, New York.

Arens T., Hettlich F., Karpfinger Ch., Kockelkorn U., Lichtenegger K., Stachel H. (2018): Mathematik, 4. Auflage, Spektrum Akademischer Verlag, Heidelberg.
Courant R. (1971a): Vorlesungen über Differential- und Integralrechnung, I. Funktionen einer Veränderlichen. Vierte Auflage, Springer Verlag, Berlin, Heidelberg, New York.
Dunham W. (1990): Journey through Genius, The Great Theorems of Mathematics. Penguin Books, New York.
Erwe F. (1962): Differential- und Integralrechnung I, II. Bibliographisches Institut, Mannheim.
Hall R. (2002): Philosophers at War: The Quarrel Between Newton and Leibniz, Cambridge University Press.
Katz V. J. (2009): A History of Mathematics, An Introduction, 3rd Edition, Addison-Wesley, Boston.
Lang S. (1986): A First Course in Calculus, Fifth Edition, Springer Verlag, New York.
Maor E. (2015): *e*: The Story of a Number, Princeton University Press, Princeton, New Jersey.
Marsden J., Weinstein A. (1985): Calculus II, Second Edition, Springer, New York.
Merzbach U.C., Boyer C.B. (2011): A History of Mathematics, Third Edition, John Wiley & Sons, Inc., Hoboken, New Jersey.
Spivak M. (2008): Calculus, Third Edition, Cambridge University Press, Cambridge.
Stillwell J. (2002): Mathematics and its History, Second Edition, Springer Verlag, New York.

Integral Calculus

4

Learning Objectives (This Chapter Provides)

- which relationship between integral and differential calculus exists
- what is meant by definite and indefinite integral
- the application of integral calculus to area calculation
- how integral calculus is used in economics ◄

4.1 The Indefinite Integral

First, we ask the question of determining a function $F(x)$, which has the following property:

$$F'(x) = f(x).$$

The task is: Given a function $f(x)$, we are looking for a function $F(x)$, which satisfies the condition $F'(x) = f(x)$.

The function $F(x)$ is called the **indefinite integral** or primitive of the function $f(x)$.

Obviously, the primitive $F(x)$ is not uniquely determined. For, if $F(x)$ satisfies the condition $F'(x) = f(x)$, then $F(x) + c$ with $c \in \mathbb{R}$ also satisfies this condition.

Examples

1. Let
$$f(x) = x^2 = F'(x),$$
then:
$$F(x) = \frac{1}{3}x^3 + c.$$

2. Let
$$f(x) = e^{2x} = F'(x),$$
then:
$$F(x) = \frac{1}{2} \cdot e^{2x} + c.$$

A notation for the primitive function of $f(x)$ is:
$$\int f(x)\,dx = F(x) + c.$$

▶ **Definition (Indefinite Integral)** $\int f(x)dx$ is referred to as the **indefinite integral** of $f(x)$, it holds:
$$\int f(x)dx = F(x) + c \quad \text{where} \quad F'(x) = f(x).$$

4.1.1 Primitives of Elementary Functions

In this section we consider primitives of some important elementary functions.

$$f(x) = x^n : \quad \int x^n\,dx = \frac{1}{n+1}x^{n+1} + c; \quad (n \neq -1). \tag{4.1}$$

$$f(x) = \frac{1}{x} : \quad \int \frac{1}{x}dx = \begin{cases} \ln x + c & x > 0 \\ \ln(-x) + c & x < 0. \end{cases} \tag{4.2}$$

$$f(x) = e^x : \quad \int e^x dx = e^x + c. \tag{4.3}$$

$$f(x) = (ax+b)^n : \quad \int (ax+b)^n dx = \frac{1}{a}\frac{(ax+b)^{n+1}}{n+1} + c; \quad n \neq -1, \tag{4.4}$$

4.1 The Indefinite Integral

with the following domains for n and x:

$$n \in \mathbb{N}; \quad a \cdot x + b \in \mathbb{R}$$
$$n \in \mathbb{Z}; \quad a \cdot x + b \neq 0$$
$$n \in \mathbb{R}; \quad a \cdot x + b > 0.$$

For the trigonometric function

$$f(x) = A \sin[b(x+c)] \qquad \text{with } A, b, c \in \mathbb{R}$$

the primitive is:

$$\int A \sin[b(x+c)]\, dx = -\frac{A}{b} \cos[b(x+c)] + d; \quad d \in \mathbb{R}. \tag{4.5}$$

The proof that $\int f(x)\, dx$ is the primitive of the function $f(x)$ is easily done by differentiating $\int f(x)\, dx + c$. Let us consider an example. Let:

$$f(x) = \frac{1}{\sqrt{2x+1}} = (2x+1)^{-\frac{1}{2}},$$

then by Eq. (4.1):

$$\int (2x+1)^{-\frac{1}{2}} dx = \sqrt{2x+1} + c.$$

If we take the derivative of the function

$$F(x) = \sqrt{2x+1},$$

we get with the help of the chain rule:

$$F'(x) = \frac{1}{2} \cdot 2(2x+1)^{-\frac{1}{2}} = \frac{1}{\sqrt{2x+1}}.$$

▶ **Note:**
Not every function $f(x)$ has a primitive $F(x)$ such that

$$F = \int f(x)\, dx.$$

4.1.2 Linearity of the indefinite integral

With the rules of differentiation, the **linearity** of the indefinite integral can be easily shown:

$$\int (a \cdot f(x) + b \cdot g(x))\, dx = a \cdot \int f(x)\, dx + b \cdot \int g(x)\, dx$$

with $a, b \in \mathbb{R}$.

Example Consider the integral

$$I = \int \left(12e^x + 109\sin x + \frac{x^2}{4} + 2\right) dx$$

$$= 12\int e^x dx + 109\int \sin x\, dx + \frac{1}{4}\int x^2 dx + \int 2 dx$$

$$= 12(e^x + c_1) - 109(\cos x + c_2) + \frac{1}{12}(x^3 + c_3) + 2(x + c_4),$$

where we use the elementary integrals from Eq. (4.1), (4.3) and (4.5).

4.2 The Definite Integral

The **definite integral** is introduced for the calculation of areas which are limited by the graph of a function $f(x)$, two parallels to the y-axis $x = a$ and $x = b$ as well as the x-axis (see Fig. 4.1).

The determination of this area is first done approximately. For this purpose, the interval $[a, b] \subset D_f$ is divided into intervals of width Δx_i.

In a first step, two values ξ_1, ξ_2 are chosen with:

$$a < \xi_1 < b; \qquad \xi_2 = b.$$

Then, for the area A, the following approximation holds:

$$A \approx \Delta x_1 \cdot f(\xi_1) + \Delta x_2 \cdot f(\xi_2)$$

with

$$\Delta x_1 = \xi_1 - a, \qquad \Delta x_2 = \xi_2 - \xi_1.$$

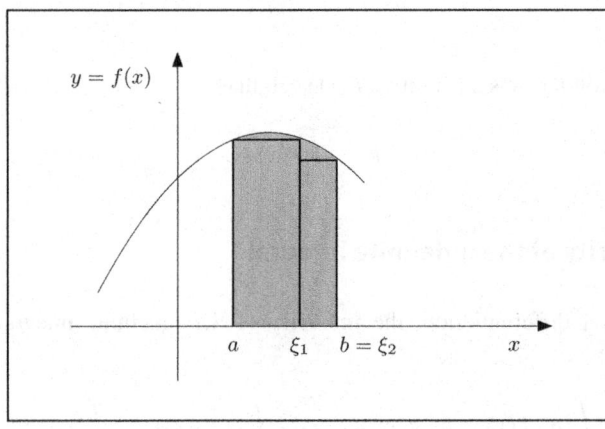

Fig. 4.1 To the calculation of areas by means of definite integrals.

4.2 The Definite Integral

The interval $[a, b]$ is now divided into n values $\xi_1, \xi_2, \ldots, \xi_n$. This improves the approximation of the area:

$$A \approx \Delta x_1 \cdot f(\xi_1) + \Delta x_2 \cdot f(\xi_2) + \ldots + \Delta x_n f(\xi_n) = \sum_{i=1}^{n} f(\xi_i) \Delta x_i$$

with

$$\Delta x_i = \xi_i - \xi_{i-1}; \quad \text{for } i > 1.$$

The choice of ξ_i is arbitrary, as long as it is:

$$\xi_1 < \xi_2 < \ldots < \xi_i < \ldots < \xi_n, \quad \text{for } i = 1, 2, 3, \ldots, n.$$

The exact area results from the limit:

$$A = \lim_{\substack{n \to \infty \\ \Delta x \to 0}} \sum_{i=1}^{n} f(\xi_i) \Delta x_i.$$

The area A depends on the limits a and b and the function $f(x)$. For this dependency we make the assumption:

$$A = F(b) - F(a)$$

with a still to be determined function $F(x)$. The relationship between the functions $F(x)$ and $f(x)$ remains open. Writing the area fractions as the difference $F(b) - F(a)$ is meaningful for visual reasons (for $b \to a$ the area goes to zero).

If $F(b) - F(a)$ is supposed to describe the desired area, then any area in the interval $[a, b]$ can be calculated between x and $x + \Delta x$ by $F(x + \Delta x) - F(x)$ (cf. Fig. 4.2).

In the interval $[x, x + \Delta x]$ the function $f(x)$ has an absolute maximum $x_{max} \in [x, x + \Delta x]$. Similarly, the function $f(x)$ has an absolute minimum x_{min} in the interval $[x, x + \Delta x]$. We now make the following estimates for the difference $F(x + \Delta x) - F(x)$:

$$\Delta x \cdot f(x_{min}) \leq F(x + \Delta x) - F(x) \leq \Delta x \cdot f(x_{max}).$$

This inequality is divided by Δx, and it follows:

$$f(x_{min}) \leq \frac{F(x + \Delta x) - F(x)}{\Delta x} \leq f(x_{max}).$$

We now carry out the limit transition $\Delta x \to 0$ or equivalently $x + \Delta x \to x$. This limit transition has the following consequences:

$$x_{min} \longrightarrow x_{max} \longrightarrow x$$

and thus:

$$f(x_{min}) \to f(x_{max}) \to f(x).$$

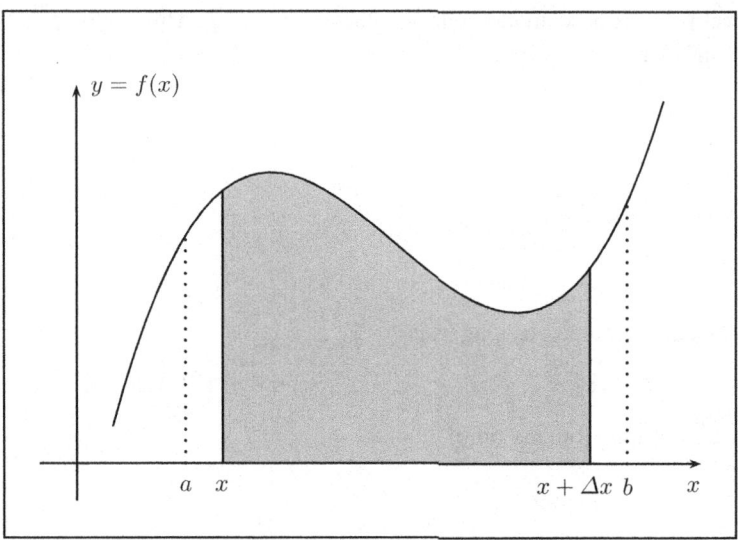

Fig. 4.2 For area determination

Therefore

$$f(x) \leq \lim_{\Delta x \to 0} \frac{F(x + \Delta x) - F(x)}{\Delta x} \leq f(x)$$

or with the definition of the derivative of a function:

$$f(x) \leq F'(x) \leq f(x).$$

It follows:

$$\boxed{F'(x) = f(x).}$$

This relationship suggests extending the notation for the indefinite integral introduced in the previous section.

▶ **Definition (Definite Integral)** The area enclosed by the function $f(x)$, the x-axis and the lines $x = a$ and $x = b$ is

$$A = \lim_{\substack{n \to \infty \\ \Delta x \to 0}} \sum_{i=1}^{n} f(\xi_i) \Delta x_i = F(b) - F(a) = \int_a^b f(x) dx. \tag{4.6}$$

$\int_a^b f(x) dx$ is called the **definite integral**.

4.2 The Definite Integral

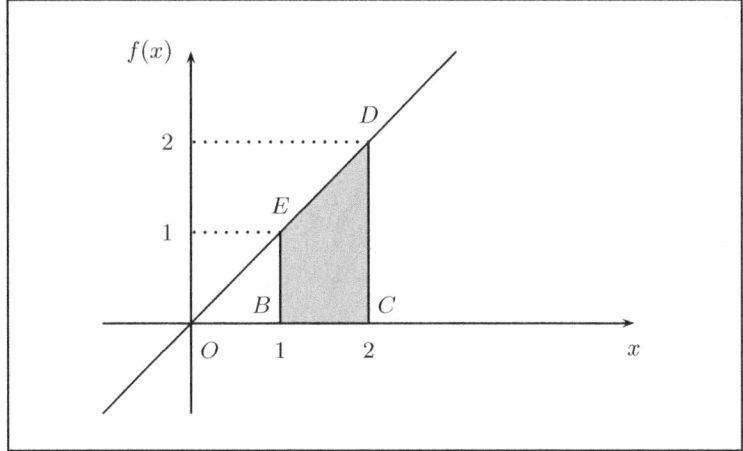

Fig. 4.3 For area calculation

The function $F(x)$ introduced for the purpose of calculating the area has the property $F'(x) = f(x)$, i.e. it is an indefinite integral of $f(x)$.

We check the above considerations using a simple example where the area can also be calculated geometrically. For this purpose, we consider the function $f(x) = x$ and calculate:

$$\int_1^2 x\,dx = \frac{1}{2}x^2 \Big|_1^2 = \frac{1}{2}2^2 - \frac{1}{2}1^2 = \frac{3}{2}.$$

From geometric considerations, it follows from Fig. 4.3:

$$F_1 = \triangle OCD = \frac{1}{2} \cdot 2 \cdot 2 = 2.$$

$$F_2 = \triangle OBE = \frac{1}{2} \cdot 1 \cdot 1 = \frac{1}{2}.$$

$$F = F_1 - F_2 = \frac{3}{2}.$$

4.2.1 Properties of the Definite Integral

The definite integral has the following properties:

1. The value of the integral does not depend on the designation of the integration variable.

$$\int_a^b f(x)\,dx = \int_a^b f(t)\,dt.$$

2. The relationship between $F(x)$ and $f(x)$ can be written as:

$$F'(x) = \frac{d}{dx} \int_a^x f(t)\,dt = f(x),$$

because:

$$\frac{d}{dx} \int_a^x f(t)\,dt = \frac{d}{dx}\left[F(x) - F(a)\right]$$
$$= \frac{dF(x)}{dx} = F'(x).$$

3. Linearity of definite integrals:
   For functions that are integrable in the interval $[a, b]$[1], we have with the constants $k_1, k_2 \in \mathbb{R}$:

$$\int_a^b \left(k_1 f(x) + k_2 g(x)\right) dx = k_1 \int_a^b f(x)\,dx + k_2 \int_a^b g(x)\,dx. \tag{4.7}$$

4. Interval additivity:
   If the function $f(x)$ is integrable in the interval $[a, b] \subset \mathbb{R}$ and $c \in [a, b]$, then:

$$\int_a^c f(x)\,dx + \int_c^b f(x)\,dx = \int_a^b f(x)\,dx. \tag{4.8}$$

5. Swapping the integration limits:
   If you swap the upper and lower integration limits, the sign of the integral is reversed:

$$\int_a^b f(x)\,dx = -\int_b^a f(x)\,dx. \tag{4.9}$$

Example Integrals play a prominent role in economics for continuous processes. As an example, we consider the continuous consumption of fuel (or energy) of a machine.

For the fuel consumption over time, we base the following functionality on it:

$$f(t) = \begin{cases} 0 & \text{for } t < t_1 \\ b \cdot e^{-at} & \text{for } t_1 \leq t \leq t_2 \\ c & \text{for } t_2 < t \leq t_3. \end{cases}$$

The graph of such a function is shown in Fig. 4.4.

The total consumption TC in the time interval $[t_1, t_3]$ is then calculated from:

$$TC = \int_{t_1}^{t_3} f(t)\,dt.$$

---

[1] As *integrable* one refers to, after Bernhard Riemann (1826–1866), functions for which the limit (4.6) exists, regardless of how the intermediate values were chosen. It can be shown that this is true for continuous functions or limited functions that only have a finite number of discontinuities.

4.2 The Definite Integral

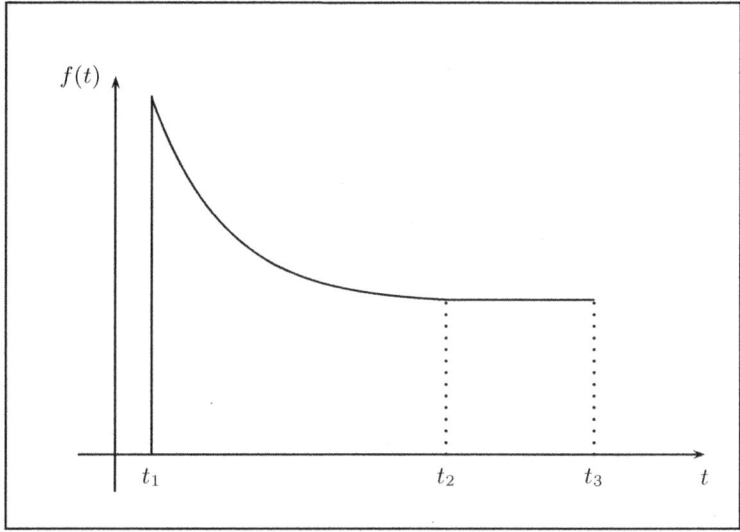

Fig. 4.4 A sectionally defined consumption function

Since the function $f(t)$ is defined sectionally, the integral TC is split into two parts:

$$TC = \int_{t_1}^{t_2} f(t)dt + \int_{t_2}^{t_3} f(t)dt.$$

It follows:

$$\begin{aligned}
TC &= \int_{t_1}^{t_2} f(t)dt + \int_{t_2}^{t_3} f(t)dt \\
&= \int_{t_1}^{t_2} b \cdot e^{-at} dt + \int_{t_2}^{t_3} c\, dt \\
&= \frac{-b}{a} \cdot e^{-at} \bigg|_{t_1}^{t_2} + c \cdot t \bigg|_{t_2}^{t_3} \\
&= \frac{-b}{a} \cdot e^{-at_2} - \frac{-b}{a} \cdot e^{-at_1} + c \cdot (t_3 - t_2) \\
&= \frac{b}{a} \cdot (e^{-at_1} - e^{-at_2}) + c \cdot (t_3 - t_2).
\end{aligned}$$

4.2.2 Value of an Integral

We consider the following function:

$$y = f(x) = \sin x.$$

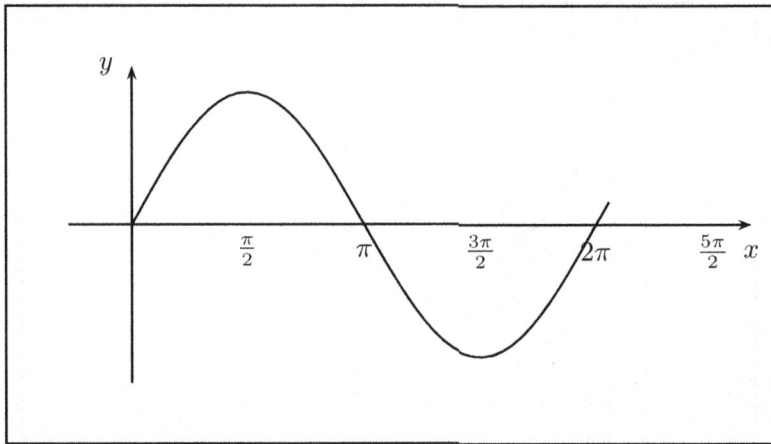

Fig. 4.5 The graph of the function $f(x) = \sin x$

The graph of the sine function is shown again in Fig. 4.5.
The primitive function of the function $f(x)$ is:

$$F(x) = -\cos x.$$

We now calculate the integral over the interval $[0, 2\pi]$:

$$\int_0^{2\pi} \sin x \, dx = -\cos x \Big|_0^{2\pi}$$
$$= -\cos 2\pi - (-\cos 0)$$
$$= -1 - (-1) = 0.$$

Obviously, this result does not correspond to the area between the curve and the x-axis. Let us consider the function $f(x) = \sin x$ on the two subintervals $[0, \pi]$ and $[\pi, 2\pi]$, so we get:

$$\int_0^{\pi} \sin x \, dx = -\cos x \Big|_0^{\pi}$$
$$= -\cos \pi - (-\cos 0)$$
$$= 1 - (-1) = 2$$

and

$$\int_{\pi}^{2\pi} \sin x \, dx = -\cos x \Big|_{\pi}^{2\pi}$$
$$= -\cos 2\pi - (-\cos \pi)$$
$$= -1 - 1 = -2.$$

4.2 The Definite Integral

The definite integral is therefore affected by a sign. For $a<b$ it holds:

$$\int_a^b f(x)\,dx > 0,$$

if $f(x)>0$ in the interval $[a, b]$ and

$$\int_a^b f(x)\,dx < 0,$$

if $f(x)<0$ in the interval $[a, b]$. Moreover, according to Eq. (4.9) the order of the integration limits is important:

$$\int_a^b f(x)\,dx = -\int_b^a f(x)\,dx.$$

From these properties of the definite integral it follows that in the area calculation the zeros of the function $f(x)$ in the interval $[a, b]$ are important. If x_0 is the only zero of the function $f(x)$ in the interval $[a, b]$, then the area A is given by:

$$A = \left|\int_a^{x_0} f(x)\,dx\right| + \left|\int_{x_0}^b f(x)\,dx\right|.$$

In the above example,

$$\int_0^{2\pi} \sin x\,dx = 0,$$

but the area that the sine function between 0 and 2π encloses with the x-axis is:

$$A = \left|\int_0^{\pi} \sin x\,dx\right| + \left|\int_{\pi}^{2\pi} \sin x\,dx\right| = 4.$$

4.2.3 Area Between Two Curves

Given two functions $f(x)$ and $g(x)$, the content of the area A between the graphs of these two functions is to be calculated (see Fig. 4.6).

This area A is the difference:

$$A = \left|\int_a^b f(x)\,dx - \int_a^b g(x)\,dx\right|$$

$$= \left|\int_a^b [f(x) - g(x)]\,dx\right|.$$

The limits a, b can be given explicitly in the form $x = a$ and $x = b$ or, as in Fig. 4.6, result from the intersection of the graphs of $f(x)$ and $g(x)$.

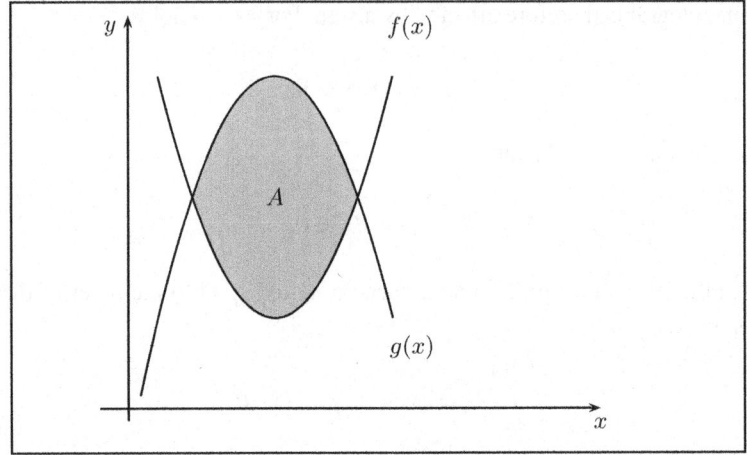

Fig. 4.6 *Area between the graphs of two functions f(x) and g(x).*

Example We consider the two functions:

$$f(x) = -x^2 + 5$$
$$g(x) = x^2 + 1$$

and calculate the content of the area enclosed between the two function graphs. The function graphs and the enclosed area are shown in Fig. 4.7.

The two intersection points of the curves result from the condition

$$f(x) \stackrel{!}{=} g(x),$$

which leads to $x_{1/2} = \pm\sqrt{2}$. The two intersection points are therefore:

$$P_1 = (-\sqrt{2}, 3) \quad \text{and} \quad P_2 = (+\sqrt{2}, 3).$$

This gives us for the enclosed area:

$$\begin{aligned}
A &= \int_{-\sqrt{2}}^{\sqrt{2}} (f(x) - g(x)) dx \\
&= \int_{-\sqrt{2}}^{\sqrt{2}} \left[(-x^2 + 5) - (x^2 + 1)\right] dx \\
&= \int_{-\sqrt{2}}^{\sqrt{2}} (-2x^2 + 4) dx \\
&= -2\frac{x^3}{3}\bigg|_{-\sqrt{2}}^{+\sqrt{2}} + 4x\bigg|_{-\sqrt{2}}^{+\sqrt{2}} \\
&= \frac{16}{3}\sqrt{2}.
\end{aligned}$$

4.2 The Definite Integral

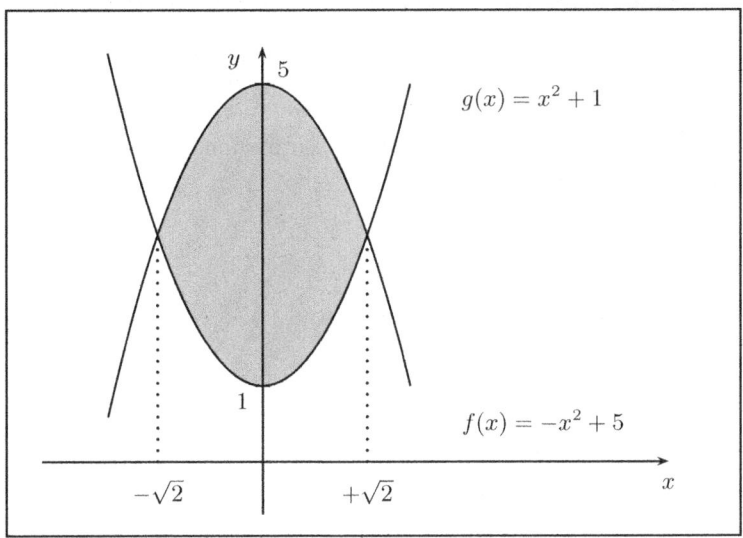

Fig. 4.7 Area between the graphs of two functions $f(x)$ and $g(x)$

4.2.4 Improper Integrals

We consider the following function:

$$g(x) = \frac{1}{x^2}; \qquad x > 0.$$

The primitive function of this function is:

$$G(x) = -\frac{1}{x} + C.$$

Then

$$\int_a^R \frac{dx}{x^2} = -\frac{1}{x}\Big|_a^R = -\frac{1}{R} + \frac{1}{a}; \quad \text{for } a > 0.$$

Let us now let the upper limit R go to $+\infty$, so the integral value apparently tends to $1/a$:

$$\lim_{R \to \infty} \int_a^R \frac{dx}{x^2} = \int_a^\infty \frac{dx}{x^2} = \frac{1}{a}, \quad \text{where } a > 0. \tag{4.10}$$

If one of the limits of a certain integral goes to infinity, one speaks of an improper integral.

▶ **Definition (Improper Integral)** The improper integral is defined by the following limit:

$$\int_a^\infty f(x)dx = \lim_{R \to \infty} \int_a^R f(x)dx. \qquad (4.11)$$

The existence of the improper integral depends on the function $f(x)$, as the following example shows. Let:

$$f(x) = \frac{1}{x}.$$

The function $f(x)$ has the primitive function:

$$F(x) = \ln x + C.$$

Then we consider the definite integral

$$\int_a^R \frac{dx}{x} = \ln x \Big|_a^R \qquad (4.12)$$
$$= \ln R - \ln a \quad \text{for } a > 0.$$

In this case, the limit $R \to \infty$ does not exist, because $\ln x$ grows beyond any limit with increasing argument. So it follows that:

$$\lim_{R \to \infty} \int_a^R \frac{dx}{x} = \infty.$$

On the other hand, if we consider the function $g(x) = \frac{1}{x^2}$ under the same aspect as in Eq. (4.10), a finite value results. This value is the area between the graph of the curve $g(x) = x^{-2}$ and the x-axis from a point $x = a$ to $x = +\infty$. The course of the two functions $f(x) = x^{-1}$ and $g(x) = x^{-2}$ for positive x values is shown in Fig. 4.8. The fact that the integral (4.12) diverges, but the integral (4.10) delivers a finite value, is related to the asymptotic behavior of the two functions $f(x)$ and $g(x)$. The function $f(x) = x^{-1}$ approaches zero asymptotically so slowly that the area under the curve always becomes larger and if x goes to infinity, it grows beyond all limits. In contrast, $g(x) = x^{-2}$ 'goes fast enough' to zero, so that a finite area under the curve results.

4.2.5 Partial Integration

We considered primitives for elementary functions in Sect. 4.1.1. In the following, we want to introduce a technique of integration that allows us to calculate primitives for other functions.

4.2 The Definite Integral

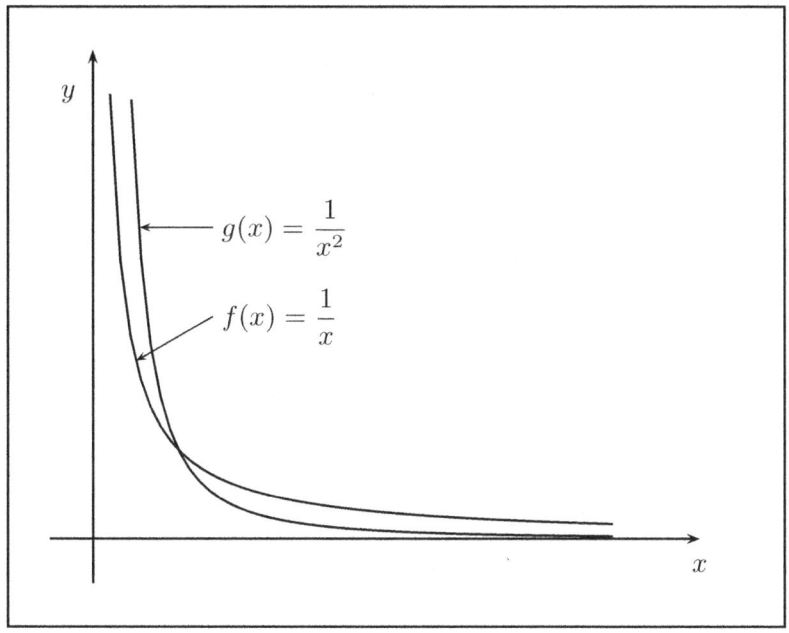

Fig. 4.8 The graphs of the two functions $f(x) = x^{-1}$ and $g(x) = x^{-2}$ for $x > 0$

We start from the formula for the differentiation of the product of two functions $f(x)$ and $g(x)$, both of which are differentiable within the same interval. Then, according to the product rule:

$$\frac{d}{dx}(f(x) \cdot g(x)) = f(x) \cdot \frac{dg(x)}{dx} + \frac{df(x)}{dx} \cdot g(x).$$

This results in integration on both sides:

$$f(x) \cdot g(x) = \int f(x) \cdot \frac{dg(x)}{dx} dx + \int \frac{df(x)}{dx} \cdot g(x) dx, \qquad (4.13)$$

or:

$$\int f(x) \cdot \frac{dg(x)}{dx} dx = f(x) \cdot g(x) - \int \frac{df(x)}{dx} \cdot g(x) dx. \qquad (4.14)$$

With the help of this formula, an integral to be calculated of the form $\int f\, g' dx$ can be reduced to another, possibly simpler integral $\int f'\, g\, dx$.

Example To calculate the integral

$$\int x \cdot e^x dx.$$

We set: $f(x) = x$ and $g'(x) = e^x$, then by applying partial integration Eq. (4.14):

$$\begin{aligned}\int x \cdot e^x dx &= x \cdot e^x - \int 1 \cdot e^x \, dx \\ &= x \cdot e^x - e^x \\ &= e^x (x-1).\end{aligned}$$

The primitive for the previously not considered function $f(x) = \ln x$ can also be determined by the method of partial integration (see Exercise 4.2).

4.2.6 Integration by Substitution

Another method—in addition to partial integration—to transform a given integral into an integral of elementary functions is the **integration by substitution**.

If $F(z)$ is a function of the variable z and $z = g(x)$, then $F(z)$ is a mediate function of x. According to the chain rule (see Gl. (3.12)) we have:

$$\frac{dF(z)}{dx} = \frac{dF(z)}{dz} \cdot \frac{dz(x)}{dx}.$$

The integration over x yields:

$$F(z) = \int \frac{dF(z)}{dz} \cdot \frac{dz(x)}{dx} dx$$

with

$$\frac{dF(z)}{dz} = f(z) = f(g(x))$$

and

$$\frac{dz(x)}{dx} = g'(x)$$

follows:

$$\int f(g(x)) \cdot g'(x) dx = \int f(z) dz. \qquad (4.15)$$

Example To calculate the integral:

$$\int x \cdot \sqrt{1-x^2}\,dx.$$

With the substitution

$$z(x) = 1 - x^2 \quad \text{we get:} \quad \frac{dz(x)}{dx} = -2x$$

and

$$dx = -\frac{dz}{2x}.$$

Then the desired integral can be written as:

$$\begin{aligned}
\int x \cdot \sqrt{1-x^2}\,dx &= \int x\sqrt{z}\,\frac{dz}{-2x} \\
&= -\frac{1}{2}\int \sqrt{z}\,dz \\
&= -\frac{1}{2}\cdot\frac{2}{3}z^{\frac{3}{2}} \\
&= -\frac{1}{3}z^{\frac{3}{2}} \\
&= -\frac{1}{3}(1-x^2)^{\frac{3}{2}}.
\end{aligned}$$

4.3 Application of Integral Calculus

4.3.1 Determination of the Economic Function from the Marginal Function

The economic function can be determined from the marginal function. As an example, we consider the marginal cost function.

From the marginal cost function $K'(x)$ (cf. Sect. 3.6.5) the cost function $K(x)$ results via the definite integral:

$$\int K'(x)\,dx = K(x) + c.$$

$K(x)$ stands for the quantity-dependent **variable** costs. The constant c is quantity-independent and thus describes the **fixed cost share** (cf. Sect. 2.2.12).

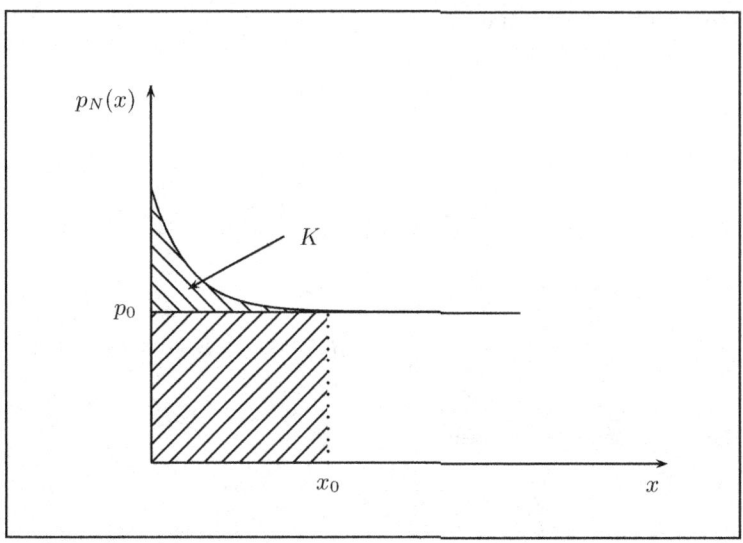

Fig. 4.9 To the term of the consumer rent

4.3.2 Consumer Rent

We consider a monotonically decreasing demand function $p(x)$, which describes the price of a product as a function of demand x. The revenue generated by the product is given by the revenue function $E_0 = p_0 \cdot x_0$ (see Sect. 2.2.12), where p_0 is the price and x_0 is the demanded quantity that results from market conditions. E_0 does not represent the maximum possible revenue from the perspective of the manufacturer, because there is a potential customer base for $x < x_0$ that would have been willing to pay a higher price. The maximum revenue results from the area under the demand function, as shown in Fig. 4.9.

$$E_{\max}(x_0) = \int_0^{x_0} p_N(x)dx.$$

The difference

$$K = E_{\max}(x_0) - E_0 = \int_0^{x_0} p_N(x)dx - p_0 x_0$$

is referred to as the **consumer surplus**.

Example Given the demand function:

$$p_N(x) = 100 \cdot e^{-0.05x}.$$

4.3 Application of Integral Calculus

The market equilibrium has been set at $x_0 = 10$ and $p_0 = 100 \cdot e^{-0.5}$. The consumer surplus is:

$$K = \int_0^{10} 100 \cdot e^{-0.05x} dx - 100 \cdot e^{-0.5} \cdot 10$$

$$= -2000 \cdot e^{-0.05x} \Big|_0^{10} - 1000 \cdot e^{-0.5}$$

$$= -2000 \cdot e^{-0.5} + 2000 - 1000 \cdot e^{-0.5}$$

$$\approx 180{,}40.$$

4.3.3 Producer Surplus

The **producer surplus** describes the additional revenue that a company makes when a product can be sold at the price p_0 (the market equilibrium), compared to a price $p_1 < p_0$, which the company had calculated for the product. In market equilibrium, the proceeds are again (Fig. 4.10)

$$E(x_0) = p_0 \cdot x_0.$$

For a monotonically increasing supply function with the price $p_A(x)$ depending on the supply x, the proceeds would be:

$$\int_0^{x_0} p_A(x) dx.$$

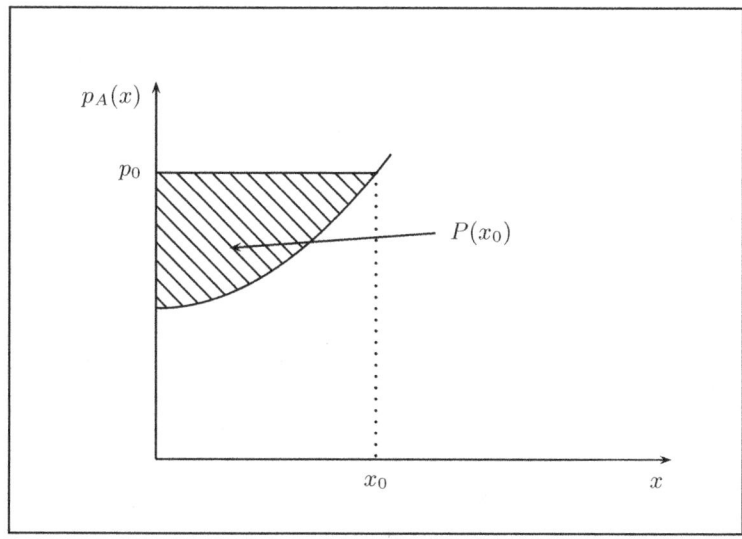

Fig. 4.10 To the concept of producer surplus

The difference

$$P(x_0) = x_0 \cdot p_0 - \int_0^{x_0} p_A(x)dx$$

results in the producer surplus.

If supply and demand functions are given, the market equilibrium is calculated from the condition

$$p_A(x_0) = p_N(x_0).$$

Then the consumer surplus and the producer surplus can be determined (see task 4.8).

4.3.4 Numerical Integration

There are a variety of functions whose primitive cannot be represented by elementary functions. A well-known example of this is the Gaussian distribution

$$f(x) = \frac{1}{\sqrt{2\pi}} e^{-\frac{x^2}{2}}.$$

The goal is to calculate the area under the Gaussian curve, i.e.

$$F[x] = \int_{-\infty}^{x} f(t)dt = \frac{1}{\sqrt{2\pi}} \int_{-\infty}^{x} e^{-t^2/2} dt.$$

The curve of the centered Gaussian distribution is shown in Fig. 4.11.
Because of the axis symmetry of the normal distribution

$$f(x) = f(-x)$$

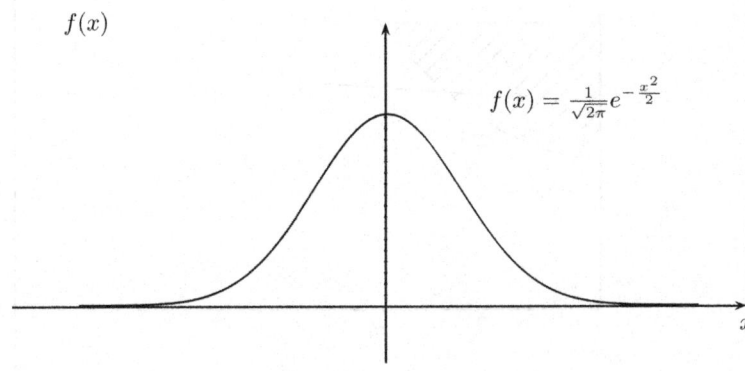

Fig. 4.11 The centered Gaussian distribution.

4.3 Application of Integral Calculus

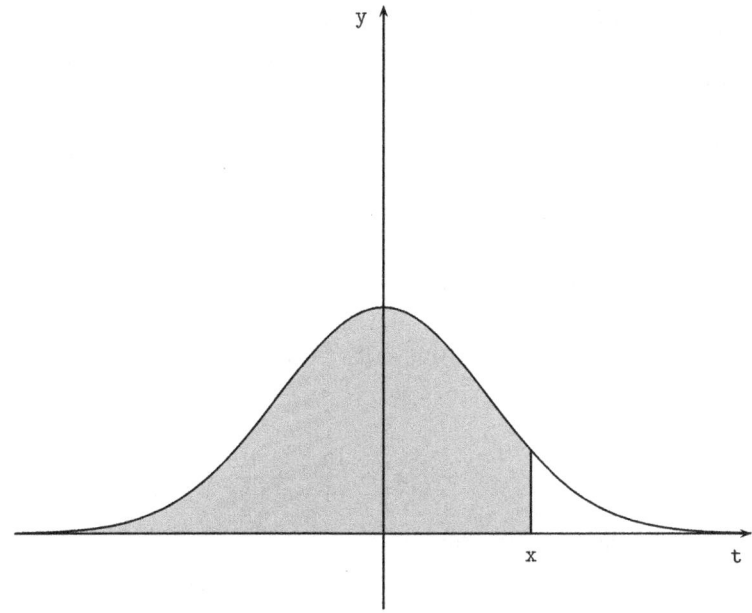

Fig. 4.12 For the integration of the Gaussian normal distribution

and the normalization

$$\int_{-\infty}^{+\infty} f(t)dt = 1$$

we can set $x \in \mathbb{R}^+$:

$$F(x) = \frac{1}{2} + \int_0^x f(t)dt = \frac{1}{2} + \frac{1}{\sqrt{2\pi}} \int_0^x e^{-t^2/2}dt.$$

In Fig. 4.12 this corresponds to the grey shaded area below the Gaussian curve.

We now develop the distribution function $f(x)$ into a Taylor series around the point $x_0 = 0$, see Eq. (3.29):

$$f(x) = \sum_{k=0}^{\infty} \frac{f^{(k)}(0)}{k!} x^k$$

$$= f(0) + \frac{f^{(1)}(0)}{1!}x + \frac{f^{(2)}(0)}{2!}x^2 + \frac{f^{(3)}(0)}{3!}x^3 + \cdots$$

The first derivatives of the function $e^{-x^2/2}$ are:

$$f(x) = e^{-x^2/2} \implies f(0) = 1$$
$$f'(x) = -xf(x) \implies f'(0) = 0$$
$$f^{(2)}(x) = (x^2 - 1)f(x) \implies f''(0) = -1$$
$$f^{(3)}(x) = (3x - x^3)f(x) \implies f^{(3)}(0) = 0$$
$$f^{(4)}(x) = (3 - 6x^2 + x^4)f(x) \implies f^{(4)}(0) = 3$$
$$f^{(5)}(x) = (-15x + 10x^3 - x^5)f(x) \implies f^{(5)}(0) = 0$$
$$f^{(6)}(x) = (-15 + 45x^2 - 15x^4 + x^6)f(x) \implies f^{(6)}(0) = -15$$
$$f^{(7)}(x) = (105x - 105x^3 + 21x^5 - x^7)f(x) \implies f^{(7)}(0) = 0$$
$$f^{(8)}(x) = (105 - 420x^2 + 210x^4 - 28x^6 + x^8)f(x) \implies f^{(8)}(0) = 105.$$

This leads to the Taylor series

$$f(x) = 1 - \frac{1}{2!}x^2 + \frac{3}{4!}x^4 - \frac{15}{6!}x^6 + \frac{105}{8!}x^8 \mp \cdots .$$

This series converges and may therefore be integrated term by term. We obtain

$$F(x) = \frac{1}{2} + \int_0^x f(t)dt$$

$$= \frac{1}{2} + \frac{1}{\sqrt{2\pi}} \int_0^x e^{-t^2/2} dt$$

$$= \frac{1}{2} + \frac{1}{\sqrt{2\pi}} \int_0^x \left(1 - \frac{1}{2!}t^2 + \frac{3}{4!} \cdot t^4 - \frac{15}{6!} \cdot t^6 + \frac{105}{8!} t^8 \mp \cdots \right)$$

$$= \frac{1}{2} + \frac{1}{\sqrt{2\pi}} \left[t - \frac{t^3}{2! \cdot 3} + \frac{3}{4! \cdot 5} t^5 - \frac{15}{6! \cdot 7} t^7 + \frac{105}{8! \cdot 9} t^9 \right.$$
$$\left. - \frac{945}{10! \cdot 11} t^{11} \pm \cdots \right]_0^x$$

$$= \frac{1}{2} + \frac{1}{\sqrt{2\pi}} \left[x - \frac{x^3}{3!} + \frac{3}{5!} x^5 - \frac{15}{7!} x^7 + \frac{105}{9!} x^9 - \frac{945}{11!} x^{11} \pm \cdots \right].$$

Numerically, for example,

$$F(1/2) = \frac{1}{2} + \frac{1}{\sqrt{2\pi}} \left(\frac{1}{2} - \frac{1}{3!} \cdot \left(\frac{1}{2}\right)^3 + \frac{3}{5!} \cdot \left(\frac{1}{2}\right)^5 - \frac{15}{7!} \left(\frac{1}{2}\right)^7 \pm \cdots \right)$$

$$= 0.69146.$$

Or

$$F(1) = \frac{1}{2} + \frac{1}{\sqrt{2\pi}} \left(1 - \frac{1}{3!} + \frac{3}{5!} - \frac{15}{7!} \pm \cdots \right)$$

$$= 0.8412.$$

For x-values like 2 or greater, higher terms of the Taylor development must be considered, since we have developed the function $f(x)$ around the point $x = 0$. For example:

$$F(2) = \frac{1}{2} + \frac{1}{\sqrt{2\pi}} \left(2 - \frac{4}{3} + \frac{4}{5} - \frac{8}{21} + \frac{4}{27} - \frac{8}{165} \pm \cdots \right)$$
$$= 0.9728$$

The tabular values of the standard normal distribution are[2]

$$F(0.500) = 0.6915, \quad F(1) = 0.8413, \quad F(2) = 0.9772.$$

4.4 Exercises

Short solutions to the following exercises can be found in the appendix.

4.1 Calculate the primitives of the following functions:

(a) $f(x) = 3 \cdot x^2$
(b) $f(x) = \sqrt{2 \cdot x}$
(c) $f(x) = (2x - 1)^2$
(d) $f(x) = e^{-ax}$
(e) $f(x) = \frac{1}{\sqrt{1-x}}$
(f) $f(x) = \sin x$

4.2 Calculate the integral

$$\int \ln x \, dx.$$

using partial integration. Note: Use:

$$\int \ln x \, dx = \int 1 \cdot \ln x \, dx.$$

Calculate

$$\int_{1/e}^{e} \ln x \, dx.$$

4.3 Consider the function

$$f(x) = \begin{cases} c & \text{if } a \leq x \leq b, (c > 0, a < b), a, b, c \in \mathbb{R}, \\ 0 & \text{else} \end{cases}$$

---

[2] See Fahrmeir et al. (2001), Appendix A.

(a) Calculate c from the requirement $\int_{-\infty}^{+\infty} f(x)dx = 1$.
(b) Sketch $f(x)$ for $a = 1, b = 3$.
(c) Sketch the indefinite integral $F(x)$ for $a = 1, b = 3$.
(d) Calculate the definite integral

$$\int_{-\infty}^{+\infty} xf(x)dx.$$

4.4 Calculate the area enclosed by the x-axis, $x = 0$, $x = 3$ and the function

$$f(x) = x^2 - 3x + 2$$

4.5 Calculate the definite integral

$$\int_0^2 \left(e^x - \frac{1}{2}\right) dx.$$

4.6 Investigate whether the following improper integral exists:

$$\int_0^\infty e^{-ax} dx; \quad a \in \mathbb{R}, a > 0.$$

4.7 Calculate the area enclosed by the two functions $f(x)$ and $g(x)$:

$$f(x) = x^3; \qquad g(x) = x.$$

4.8 Given the demand function

$$p_N(x) = 25 - x^2$$

and the supply function

$$p_A(x) = 10 + 2x.$$

Calculate the market equilibrium, the consumer surplus and the producer surplus.

4.9 Calculate the integral

$$I = \int e^{-x} \sin x \, dx.$$

4.10

(a) Determine the indefinite integral

$$I_1 = \int x \cdot e^{-x^2} dx.$$

(b) Does the improper integral exist?

$$I_2 = \int_0^\infty x \cdot e^{-x^2} dx.$$

4.4 Exercises

4.11 Determine the indefinite integral

$$I = \int \frac{x^3}{x^4 - 5} dx.$$

4.12 The capital flow—that is the change in capital over time—is modeled by the function

$$f(t) = -t^3 + 36t, \quad t \geq 0.$$

The variable t is time in months, $f(t)$ the capital flow in 10^3 € per month. The initial capital is $K_0 = 576{,}000$ €.

(a) When is the capital growth maximal?
(b) At what time is the capital the largest? How large is the capital at this time?
(c) When is the initial capital K_0 used up?

4.13 A publisher has at the time $t = 0$ a million subscribers. The newly added subscribers are modeled by the time-dependent function $n(t)$, the canceled subscribers by the function $k(t)$, each in one hundred thousand subscribers per year. The change rate is therefore $n(t)-k(t)$. It is

$$n(t) = \frac{5}{t+2} \quad \text{for } t \geq 0, \qquad k(t) = -\frac{3}{t+1} + 4 \quad \text{for } t \geq 0.$$

When are there the most subscribers (solution $t = 0.5$)? How many are there?

4.14 Calculate the following definite integrals:

$$I_1 = \int_1^2 \frac{dx}{x^2},$$

$$I_2 = \int_{-3}^{-1} \frac{1+x}{x^3} dx,$$

$$I_3 = \int_{1/2}^{3} \frac{x^2 - 2}{x^4} dx,$$

$$I_4 = 2 \cdot \int_2^1 \frac{1 - x^4}{x^2} dx.$$

4.15 Calculate the definite integral

$$\int_0^a \frac{2}{1 - x^2} dx.$$

What condition must a fulfill so that the integral exists?

Linear Algebra 5

Learning Objectives (This Chapter Covers)

- a basic introduction to vector and matrix calculus
- which operations for vectors and matrices are of interest
- the treatment and solution of linear equation systems
- how economic problems can be solved using linear algebra ◄

5.1 Vectors

The central concept of linear algebra are vectors and operations that can be performed with these objects.[1]

5.1.1 Definition of Vectors

The representation of points in a plane provides an intuitive picture of vectors. In geometry, one speaks of the **position vector** of a point $P(x_1, x_2)$.

Figure 5.1 shows a two-dimensional Cartesian coordinate system with the origin O and the unit vectors

$$\mathbf{e}_1 = \begin{pmatrix} 1 \\ 0 \end{pmatrix} \text{ and } \mathbf{e}_2 = \begin{pmatrix} 0 \\ 1 \end{pmatrix}.$$

---

[1] Deeper considerations of linear algebra can be found, for example, in Goebbels and Ritter (2011). Chap. 3, Fischer and Springborn (2020) or S. Lang (1987). The recommended book by Deisenroth et al. (2020) discusses current applications of linear algebra in data science.

© The Author(s), under exclusive license to Springer-Verlag GmbH, DE, part of Springer Nature 2023
T. Holey and A. Wiedemann, *Analysis and Linear Algebra*,
https://doi.org/10.1007/978-3-662-66247-2_5

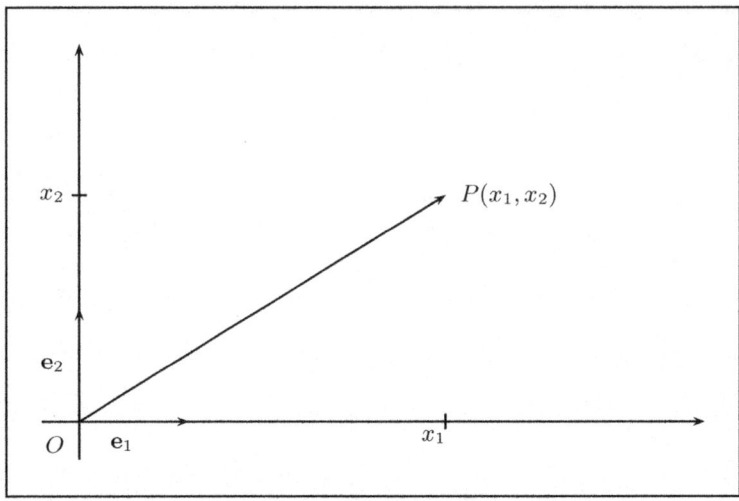

Fig. 5.1 To the concept of the position vector

The position vector of a point with the coordinates x_1 and x_2 is then:

$$\mathbf{OP} = x_1 \cdot \mathbf{e_1} + x_2 \cdot \mathbf{e_2}$$
$$= x_1 \begin{pmatrix} 1 \\ 0 \end{pmatrix} + x_2 \begin{pmatrix} 0 \\ 1 \end{pmatrix}$$
$$= \begin{pmatrix} x_1 \\ x_2 \end{pmatrix}.$$

In contrast to **scalar quantities** such as time or temperature, which can be described by a real number, vectors are characterized by the fact that several components are required to specify a vector. For three-dimensional space, this concept can be generalized:

$$\mathbf{OP} = \begin{pmatrix} x_1 \\ x_2 \\ x_3 \end{pmatrix}.$$

The properties of these position vectors are determined by the **magnitude** of the vector (i.e. its length) and the **direction**. The magnitude of the vector results from the Pythagorean theorem to

$$l = \sqrt{x_1^2 + x_2^2 + x_3^2}.$$

In a generalized view, we define a vector in the following form:

5.1 Vectors

▶ **Definition (Vector)**

A vector is an ordered column of n numbers. With vertical arrangement of the numbers, one speaks of **column vectors**

$$\mathbf{x} = \begin{pmatrix} x_1 \\ x_2 \\ x_3 \\ \vdots \\ x_n \end{pmatrix}$$

and with horizontal arrangement of **row vectors**

$$\mathbf{y}^\top = (y_1, y_2, y_3, \ldots, y_n).$$

Here, in analogy to Sect. 5.2, the notation $\mathbf{y}^\top$ has prevailed, where ⊤ stands for *transposed*.[2]

Examples

1. In industrial production, n-dimensional vectors are used to describe the daily production of n machines:

$$\mathbf{x} = \begin{pmatrix} x_1 \\ x_2 \\ x_3 \\ \vdots \\ x_n \end{pmatrix}$$

or

$$\mathbf{x}^\top = (x_1, x_2, \ldots, x_n).$$

2. In sales, the prices of m products are captured by a vector with m components:

$$\mathbf{p} = \begin{pmatrix} p_1 \\ p_2 \\ p_3 \\ \vdots \\ p_m \end{pmatrix} \quad \text{or} \quad \mathbf{p}^\top = (p_1, p_2, \ldots, p_m).$$

---

[2] Some properties of geometric vectors such as representation with a length and a direction as well as transformation behavior cannot be transferred to the general consideration of vectors as ordered columns of numbers. But these considerations do not play a role in economic applications and are not pursued here.

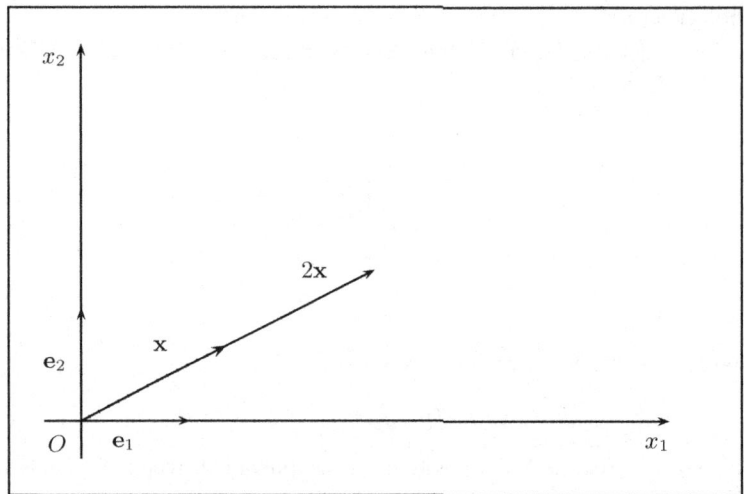

Fig. 5.2 Multiplication of a vector by a scalar

5.1.2 The Linear Combination of Vectors

Vectors can be multiplied by a scalar, i.e. by a real number:

$$s \cdot \mathbf{x} = \begin{pmatrix} sx_1 \\ sx_2 \\ \vdots \\ sx_n \end{pmatrix}, \quad s \in \mathbb{R}.$$

In the geometric visualization, multiplication by a scalar is merely a multiplication[3] of the original vector (see Fig. 5.2).

The addition of two vectors:

$$\mathbf{a} + \mathbf{b} = \begin{pmatrix} a_1 + b_1 \\ a_2 + b_2 \\ \vdots \\ a_n + b_n \end{pmatrix}$$

can also be geometrically illustrated (see Fig. 5.3).

---

[3] More precisely, a stretching.

5.1 Vectors

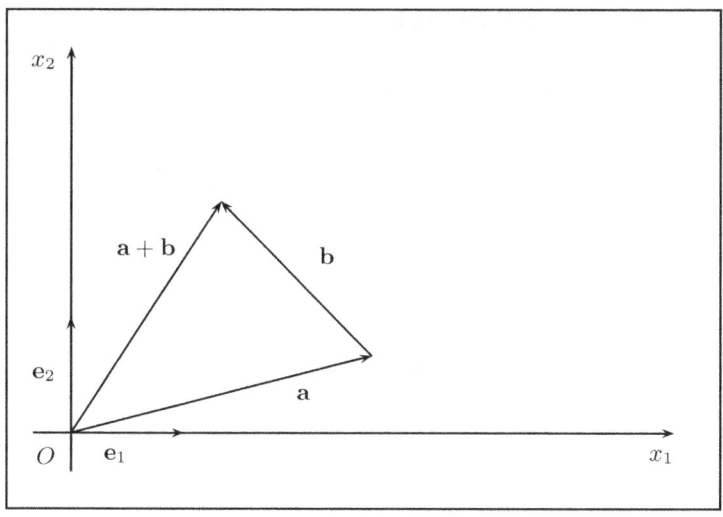

Fig. 5.3 Addition of two vectors **a** + **b**

A **linear combination** of vectors $\mathbf{a}_1, \mathbf{a}_2, \ldots, \mathbf{a}_n$ is defined as vector:

$$\mathbf{x} = x_1\mathbf{a}_1 + x_2\mathbf{a}_2 + \cdots + x_n\mathbf{a}_n = \sum_{i=1}^{n} x_i \mathbf{a}_i$$

with n real coefficients x_i.

In the following we will need the concept of **linear independence of vectors**.

▶ **Definition (Linear Independence)** The n vectors $\mathbf{a}_i, i = 1, 2, \ldots, n$ are called **linearly independent**, if the null vector (that is the vector where all components are zero) can only be represented in exactly one unique way, in which all factors are $x_i = 0$. The equation

$$\mathbf{0} = \sum_{i=1}^{n} x_i \mathbf{a}_i$$

only has the unique solution $x_1 = x_2 = \cdots = x_n = 0$.

Intuitively, linear independence of vectors means that none of the vectors can be represented by a linear combination of the other vectors.

5.1.3 Scalar Product of Two Vectors

We formally define the **scalar product** of two vectors $\mathbf{a}^\top$ and $\mathbf{b}$ in the form:

$$\mathbf{a}^\top \cdot \mathbf{b} = \begin{pmatrix} a_1 & a_2 & \ldots & a_n \end{pmatrix} \cdot \begin{pmatrix} b_1 \\ b_2 \\ \vdots \\ b_n \end{pmatrix}$$

$$= a_1 b_1 + a_2 b_2 + \cdots + a_n b_n$$

$$= \sum_{i=1}^{n} a_i b_i \in \mathbb{R}.$$

Example Consider the sales volumes

$$\mathbf{x}^\top = \begin{pmatrix} x_1 & x_2 & x_3 \end{pmatrix} = (2, 5, 7)$$

of three products, we can summarize the prices of these products in a vector

$$\mathbf{p} = \begin{pmatrix} p_1 \\ p_2 \\ p_3 \end{pmatrix} = \begin{pmatrix} 5 \\ 5 \\ 1 \end{pmatrix}.$$

The scalar product $\mathbf{x}^\top \cdot \mathbf{p}$ provides the revenue generated with all products together:

$$U = \mathbf{x}^\top \cdot \mathbf{p} = (2, 5, 2) \cdot \begin{pmatrix} 5 \\ 5 \\ 1 \end{pmatrix} = 10 + 25 + 2 = 37.$$

Two vectors whose scalar product is zero are called **orthogonal**. Consider the two vectors

$$\mathbf{a} = \begin{pmatrix} 1 \\ 1 \end{pmatrix}, \mathbf{b} = \begin{pmatrix} -1 \\ 1 \end{pmatrix}, \quad \mathbf{a}, \mathbf{b} \in \mathbb{R}^2.$$

Then

$$\mathbf{a}^\top \cdot \mathbf{b} = (1, 1) \cdot \begin{pmatrix} -1 \\ 1 \end{pmatrix} = -1 + 1 = 0.$$

Orthogonal vectors are perpendicular to each other, as shown in Fig. 5.4.

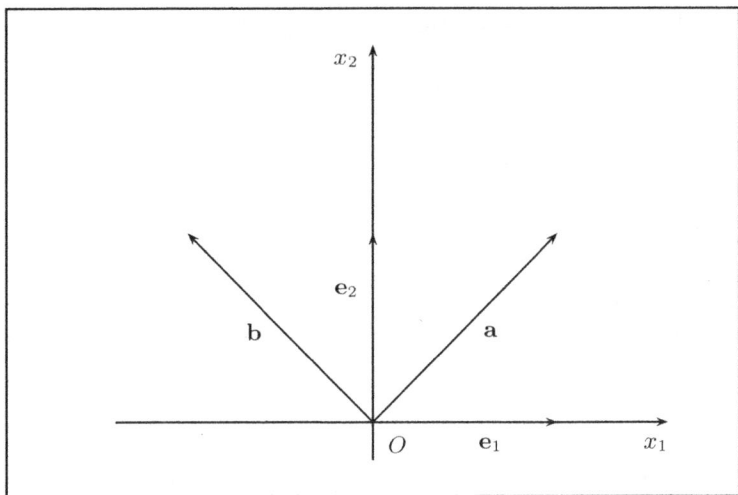

Fig. 5.4 Orthogonality of two vectors **a** and **b**

5.2 Matrices

5.2.1 Definition of a Matrix

The definition of vectors as ordered number columns can be generalized. We now consider the representation of data that is in tabular form.

▶ **Definition (Matrix)** A **matrix** is a schematic representation of m rows and n columns

$$\mathbf{A} = \begin{pmatrix} a_{11} & a_{12} & \cdots & a_{1n} \\ a_{21} & a_{22} & \cdots & a_{2n} \\ a_{31} & a_{32} & \cdots & a_{3n} \\ \vdots & \vdots & \vdots & \vdots \\ a_{m1} & a_{m2} & \cdots & a_{mn} \end{pmatrix}$$

A matrix element a_{ij} is defined by the row index i ($i = 1, 2, \ldots, m$) and the column index j with ($j = 1, 2, \ldots, n$).

Example As an example of the application of matrices in economics, we consider the raw material consumption coefficients. They indicate how many units of different raw materials are needed to produce one unit of different products.

	Product 1	Product 2	Product 3
Raw material 1	1	2	3
Raw material 2	0	3	1

This numerical scheme leads to the matrix:

$$\mathbf{R} = \begin{pmatrix} 1 & 2 & 3 \\ 0 & 3 & 1 \end{pmatrix}.$$

The matrix entry $r_{23} = 1$ means: For each unit of product 3, one unit of raw material 2 is needed.

With the definition of a matrix, further basic concepts for matrices can be introduced:

- The **square matrix**
  If $m = n$, one speaks of a *square matrix*. The elements

  $$a_{11}, a_{22}, \ldots, a_{nn}$$

  are then called *diagonal elements*.

- The **transposed matrix**
  The transposition of rows and columns of a matrix is called **transposition**. Given a matrix:

  $$\mathbf{A} = (a_{ij}) \quad \text{where} \quad i = 1, 2, \ldots, m; \ j = 1, 2, \ldots, n,$$

  then the transposed matrix is given by

  $$\mathbf{A}^\top = (a_{ji}) \quad \text{where} \quad i = 1, 2, \ldots, m; \ j = 1, 2, \ldots, n.$$

  Example Given the 2×3 matrix:

  $$\mathbf{A} = \begin{pmatrix} 2 & 3 & 5 \\ 1 & 4 & 7 \end{pmatrix},$$

  then the transposed matrix is a 3×2 matrix with

  $$\mathbf{A}^\top = \begin{pmatrix} 2 & 1 \\ 3 & 4 \\ 5 & 7 \end{pmatrix}.$$

  It holds:

  $$(\mathbf{A}^\top)^\top = \mathbf{A}.$$

  Due to this property, one says that the transposition is idempotent.

- **Symmetric Matrices**
  For square matrices, it may happen that

  $$\mathbf{A}^\top = \mathbf{A},$$

  in this case $\mathbf{A}$ is called a **symmetric matrix**.
  Example The matrix

$$A = \begin{pmatrix} 2 & 3 & 5 \\ 3 & 4 & 7 \\ 5 & 7 & 9 \end{pmatrix}$$

is a symmetric 3×3 matrix, because it is $a_{ij} = a_{ji}$.

- **Antisymmetric Matrices**

  Again for square matrices, it may happen that

  $$A^\top = -A,$$

  then A is antisymmetric. For the matrix elements, this means:

  $$a_{ij} = -a_{ji} \text{ and } a_{ii} = 0.$$

  Examples The matrix

  $$A = \begin{pmatrix} 0 & -1 \\ 1 & 0 \end{pmatrix}$$

  is antisymmetric, because it has the property $A^\top = -A$.

- **Vectors**

  Vectors are special matrices. A vector of the form:

  $$a = \begin{pmatrix} a_1 \\ a_2 \\ \vdots \\ a_n \end{pmatrix}$$

  is a matrix with n rows and one column, that is, an $n \times 1$ matrix. A row vector

  $$a^\top = (a_1, a_2, \ldots, a_n)$$

  is then a matrix with one row and n columns, that is, a $1 \times n$ matrix.

- The **Zero matrix**

  The *Zero matrix* is defined as the $m \times n$ matrix with $a_{ij} = 0$ for all $i = 1, 2, \ldots, m$; $j = 1, 2, \ldots, n$. For the Zero matrix we write: $0_{m \times n}$.

- The **identity matrix**

  The *identity matrix* is a symmetric $n \times n$ matrix with

  $$\mathbb{1}_{n \times n} = \begin{pmatrix} 1 & 0 & 0 & \cdots & 0 \\ 0 & 1 & 0 & \cdots & 0 \\ \vdots & \vdots & \vdots & & \\ 0 & 0 & 0 & \cdots & 1 \end{pmatrix} = E.$$

  This matrix is sometimes also written in the form:

$$\mathbb{1}_{n\times n} = (\delta_{ij}); \quad i,j = 1,2,\ldots,n$$

with the Kronecker delta symbol:

$$\delta_{ij} = \begin{cases} 1 & \text{if } i=j; \quad i,j = 1,2,\ldots,n \\ 0 & \text{else} \end{cases}.$$

5.2.2 Addition of Matrices

▶ **Definition (Addition of Matrices)** Given two $m \times n$ matrices **A** and **B**, then the sum of these two matrices is again an $m \times n$ matrix:

$$\mathbf{C} = \mathbf{A} + \mathbf{B}$$

with the component-wise addition:

$$c_{ij} = a_{ij} + b_{ij}, \quad \text{where } i = 1,2,\ldots,m; j = 1,2,\ldots,n.$$

Example For the production of the products P_1 and P_2 the order quantity of the customers K_1, K_2, K_3 is decisive. For the production planning the production from two quarters Q_1, Q_2 is considered.

	Product 1	Product 2	Product 1	Product 2
Customer 1	1	2	2	1
Customer 2	0	1	0	1
Customer 3	3	2	0	2
	Q_1		Q_2	

Thus the matrix

$$\mathbf{B}_1 = \begin{pmatrix} 1 & 2 \\ 0 & 1 \\ 3 & 2 \end{pmatrix}$$

stands for the order in the first quarter Q_1 and the matrix

$$\mathbf{B}_2 = \begin{pmatrix} 2 & 1 \\ 0 & 1 \\ 0 & 2 \end{pmatrix}$$

for the order in the quarter Q_2. The total order results in:

$$\mathbf{B} = \mathbf{B}_1 + \mathbf{B}_2 = \begin{pmatrix} 3 & 3 \\ 0 & 2 \\ 3 & 4 \end{pmatrix}.$$

5.2 Matrices

The following laws apply to the addition of matrices:

1. **Associative law:**
$$(A + B) + C = A + (B + C).$$

2. **Commutative law:**
$$A + B = B + A.$$

3. **Linearity of transposition:**
$$(A + B)^\top = A^\top + B^\top.$$

5.2.3 Multiplication by a Scalar

▶ **Definition (Multiplication of a Matrix with a Scalar)** Let $s \in \mathbb{R}$ be a real number and A an $m \times n$-matrix. Then multiplication with the scalar quantity s is component-wise defined by:
$$s \cdot A = (s \cdot a_{ij}) \quad \text{for} \quad i = 1, 2, \ldots, m; \quad j = 1, 2, \ldots n.$$

The following rules apply to scalar multiplication:

1. **Commutativity:**
$$s \cdot A = A \cdot s.$$

2. **Associativity:**
$$(s_1 \cdot s_2) \cdot A = s_1 \cdot (s_2 \cdot A).$$

3. **Distributivity with respect to matrix addition:**
$$s \cdot (A + B) = s \cdot A + s \cdot B.$$

4. **Distributivity with respect to scalar addition:**
$$(s_1 + s_2) \cdot A = s_1 \cdot A + s_2 \cdot A.$$

5.2.4 Matrix Multiplication

▶ **Definition (Matrix Multiplication)** The product of two matrices A and B with
$$A = (a_{ij}); \quad i = 1, 2, \ldots, m; \quad j = 1, 2, \ldots n$$

and

$$\mathbf{B} = (b_{jk}); \quad j = 1, 2, \ldots, n; \quad k = 1, 2, \ldots l$$

is defined as

$$\mathbf{C} = \mathbf{A} \cdot \mathbf{B}$$

with:

$$c_{ik} = \sum_{j=1}^{n} a_{ij} \cdot b_{jk} \tag{5.1}$$

with $i = 1, 2, \ldots, m$ and $k = 1, 2, \ldots, l$. The index $j = 1, 2, \ldots, n$ is summed over.

It should be emphasized at this point that the matrix product (5.1) is only defined if the number of columns of matrix $\mathbf{A}$ agrees with the number of rows of matrix $\mathbf{B}$.

Examples
1. We consider the 2×2 matrix $\mathbf{A}$ with:

$$\mathbf{A} = \begin{pmatrix} a_{11} & a_{12} \\ a_{21} & a_{22} \end{pmatrix}$$

and the 2×3 matrix $\mathbf{B}$ with

$$\mathbf{B} = \begin{pmatrix} b_{11} & b_{12} & b_{13} \\ b_{21} & b_{22} & b_{23} \end{pmatrix}.$$

Then the product of these two matrices is given by the 2×3 matrix $\mathbf{C}$ with components

$$c_{ik} = \sum_{j=1}^{2} a_{ij} \cdot b_{jk}$$

and:

$$c_{11} = a_{11} \cdot b_{11} + a_{12} \cdot b_{21}$$
$$c_{21} = a_{21} \cdot b_{11} + a_{22} \cdot b_{21}$$
$$c_{12} = a_{11} \cdot b_{12} + a_{12} \cdot b_{22}$$
$$c_{22} = a_{21} \cdot b_{12} + a_{22} \cdot b_{22}$$
$$c_{13} = a_{11} \cdot b_{13} + a_{12} \cdot b_{23}$$
$$c_{23} = a_{21} \cdot b_{13} + a_{22} \cdot b_{23}.$$

So the product matrix is:

$$\mathbf{C} = \begin{pmatrix} c_{11} & c_{12} & c_{13} \\ c_{21} & c_{22} & c_{23} \end{pmatrix}.$$

5.2 Matrices

The example shows that matrix multiplication can be considered as a scalar product of row and column vectors. The row vector from the i-th row of matrix $\mathbf{A}$ is scalar multiplied with the column vector of the k-th column of $\mathbf{B}$ and provides the element c_{jk} of the product matrix. Consider for example:

$$c_{13} = (a_{11}\ a_{12}) \cdot \begin{pmatrix} b_{13} \\ b_{23} \end{pmatrix} = a_{11}b_{13} + a_{12}b_{23}.$$

2. If

$$\mathbf{A} = \begin{pmatrix} 2 & 3 \\ 4 & 3 \end{pmatrix} \text{ and } \mathbf{B} = \begin{pmatrix} 4 & 3 \\ 2 & 1 \end{pmatrix},$$

then the product matrix $\mathbf{C}$ is explicitly given by:

$$\mathbf{C} = \mathbf{A} \cdot \mathbf{B}$$

$$= \begin{pmatrix} 2 & 3 \\ 4 & 3 \end{pmatrix} \cdot \begin{pmatrix} 4 & 3 \\ 2 & 1 \end{pmatrix}$$

$$= \begin{pmatrix} 2 \cdot 4 + 3 \cdot 2 & 2 \cdot 3 + 3 \cdot 1 \\ 4 \cdot 4 + 3 \cdot 2 & 4 \cdot 3 + 3 \cdot 1 \end{pmatrix}$$

$$= \begin{pmatrix} 14 & 9 \\ 22 & 15 \end{pmatrix}.$$

Why the definition of matrix multiplication is meaningful in this way is shown by the following consideration of an example for a multi-stage production.

A company produces two types of final products E_1 and E_2. These final products are made from three different types of intermediate products. We refer to these as Z_1, Z_2, Z_3. The intermediate products themselves are again produced from four different raw materials $R_1,\ldots, R_4$. Different amounts of the different raw materials are required for each intermediate product Z_1, Z_2, Z_3. The raw material consumption coefficients could, for example, look like this:

Raw material	Z_1	Z_2	Z_3
R_1	4	3	3
R_2	2	4	6
R_3	1	7	4
R_4	3	3	0.

We summarize this assignment in a 4×3 matrix $\mathbf{A}$:

$$\mathbf{A} = \begin{pmatrix} 4 & 3 & 3 \\ 2 & 4 & 6 \\ 1 & 7 & 4 \\ 3 & 3 & 0 \end{pmatrix}.$$

An analogous scheme can also be set up for the 2nd production stage with the production coefficients of the intermediate products:

Intermediate Product	E_1 E_2
Z_1	6 5
Z_2	4 3
Z_3	1 2.

This means that 6 units of the intermediate product Z_1, 4 units of the intermediate product Z_2 and 1 unit of the intermediate product Z_3 are required to produce one unit of the final product E_1. This assignment is written in a 3×2 matrix $\mathbf{B}$:

$$\mathbf{B} = \begin{pmatrix} 6 & 5 \\ 4 & 3 \\ 1 & 2 \end{pmatrix}.$$

The amount of raw materials needed is solely determined by the number of finished products to be produced.

If we form the matrix product in the form defined in Eq. (5.1),

$$\mathbf{C} = \mathbf{A} \cdot \mathbf{B}$$

$$= \begin{pmatrix} 4 & 3 & 3 \\ 2 & 4 & 6 \\ 1 & 7 & 4 \\ 3 & 3 & 0 \end{pmatrix} \cdot \begin{pmatrix} 6 & 5 \\ 4 & 3 \\ 1 & 2 \end{pmatrix}$$

$$= \begin{pmatrix} 39 & 35 \\ 34 & 34 \\ 38 & 34 \\ 30 & 24 \end{pmatrix},$$

we obtain from the matrix product the production coefficients of the raw materials with respect to the final products.

Raw material	E_1 E_2
R_1	39 35
R_2	34 34
R_3	38 34
R_4	30 24.

Let us consider the first element in the first row and first column: For the production of one unit of E_1, 39 units of raw material R_1 are needed. These consist of the units of intermediate products required for the production of E_1 and for each intermediate product a certain amount of R_1 is required again.

5.2.5 Calculation Rules of the Matrix Product

In this section we give some rules which are helpful for calculating the product of matrices. The proofs follow—as shown exemplarily for the associativity law—from the careful consideration of the indices.

1. **Associativity:**

$$(A \cdot B) \cdot C = A \cdot (B \cdot C). \tag{5.2}$$

Proof Let $D = (A \cdot B) \cdot C$ and $D' = A \cdot (B \cdot C)$, then:

$$\begin{aligned}
d_{ij} &= ((A \cdot B) \cdot C)_{ij} \\
&= \sum_{k=1}^{m} (A \cdot B)_{ik} \cdot c_{kj} \\
&= \sum_{k=1}^{m} (\sum_{l=1}^{n} a_{il} \cdot b_{lk}) \cdot c_{kj} \\
&= \sum_{l=1}^{n} a_{il} \cdot (\sum_{k=1}^{m} b_{lk} \cdot c_{kj}) \\
&= \sum_{l=1}^{n} a_{il} \cdot (B \cdot C)_{lj} \\
&= (A \cdot (B \cdot C))_{ij} \\
&= (d')_{ij}.
\end{aligned}$$

Since we are considering finite matrices here, the exchange of sums is allowed.

2. **Associativity of multiplication with scalars:**
   For all $s \in \mathbb{R}$ it holds:

$$s \cdot (A \cdot B) = (s \cdot A) \cdot B. \tag{5.3}$$

3. **Distributive law:**

$$A \cdot (B + C) = A \cdot B + A \cdot C. \tag{5.4}$$

4. Multiplication with the **identity matrix:**
   Let A be a $n \times n$ matrix and $\mathbb{1}_{n \times n}$ the $n \times n$ identity matrix. Then it holds:

$$A \cdot \mathbb{1}_{n \times n} = \mathbb{1}_{n \times n} \cdot A = A. \tag{5.5}$$

5. Multiplication with the **zero matrix**:

The multiplication of any matrix **A** with the zero matrix results in the zero matrix:

$$\mathbf{A} \cdot \mathbf{0}_{n \times n} = \mathbf{0}_{n \times n} \cdot \mathbf{A} = \mathbf{0}. \tag{5.6}$$

▶ **Note:**

The matrix multiplication has the following property: The product of two matrices can result in the zero matrix, even though both matrices are different from the zero matrix.

Example If

$$\mathbf{A} = \begin{pmatrix} 1 & 1 & 1 \\ 2 & 2 & 2 \end{pmatrix} \quad \text{and} \quad \mathbf{B} = \begin{pmatrix} 1 & -1 \\ 1 & -1 \\ -2 & 2 \end{pmatrix},$$

then:

$$\mathbf{A} \cdot \mathbf{B} = \begin{pmatrix} 1 & 1 & 1 \\ 2 & 2 & 2 \end{pmatrix} \cdot \begin{pmatrix} 1 & -1 \\ 1 & -1 \\ -2 & 2 \end{pmatrix} = \begin{pmatrix} 0 & 0 \\ 0 & 0 \end{pmatrix} = \mathbf{0}_{2 \times 2}.$$

6. **Transposition:**

It holds:

$$(\mathbf{A} \cdot \mathbf{B})^\top = \mathbf{B}^\top \cdot \mathbf{A}^\top. \tag{5.7}$$

Example As an example to illustrate the property (5.7) we consider the two matrices:

$$\mathbf{A} = \begin{pmatrix} 1 & 5 \\ 2 & 3 \end{pmatrix} \quad \text{and} \quad \mathbf{B} = \begin{pmatrix} 2 & 3 \\ 4 & 5 \end{pmatrix}.$$

The transposed matrices are:

$$\mathbf{A}^\top = \begin{pmatrix} 1 & 2 \\ 5 & 3 \end{pmatrix} \quad \text{and} \quad \mathbf{B}^\top = \begin{pmatrix} 2 & 4 \\ 3 & 5 \end{pmatrix}.$$

Then the product is:

$$\mathbf{A} \cdot \mathbf{B} = \begin{pmatrix} 22 & 28 \\ 16 & 21 \end{pmatrix}.$$

and the transposition of this product is:

$$(\mathbf{A} \cdot \mathbf{B})^\top = \begin{pmatrix} 22 & 16 \\ 28 & 21 \end{pmatrix}.$$

5.2 Matrices

On the other hand we have:

$$\mathbf{B}^\top \cdot \mathbf{A}^\top = \begin{pmatrix} 2 & 4 \\ 3 & 5 \end{pmatrix} \cdot \begin{pmatrix} 1 & 2 \\ 5 & 3 \end{pmatrix}$$
$$= \begin{pmatrix} 22 & 16 \\ 28 & 21 \end{pmatrix}$$
$$= (\mathbf{A} \cdot \mathbf{B})^\top.$$

7. **Non-commutativity:**
   Matrix multiplication is in general not commutative, that is:

$$\mathbf{A} \cdot \mathbf{B} \neq \mathbf{B} \cdot \mathbf{A} \tag{5.8}$$

as a simple example shows:

$$\mathbf{A} = \begin{pmatrix} 1 & 2 \\ 2 & 3 \end{pmatrix} \text{ and } \mathbf{B} = \begin{pmatrix} 2 & 0 \\ 1 & 1 \end{pmatrix},$$

then is:

$$\mathbf{A} \cdot \mathbf{B} = \begin{pmatrix} 4 & 2 \\ 7 & 3 \end{pmatrix}$$

and

$$\mathbf{B} \cdot \mathbf{A} = \begin{pmatrix} 2 & 2 \\ 7 & 3 \end{pmatrix}.$$

5.2.6 Inverse Matrix

Multiplicative inverse elements x^{-1} usually have the property

$$x \cdot x^{-1} = 1.$$

The product of an element x with its inverse element x^{-1} therefore leads to the neutral element of multiplication, which has the property

$$a \cdot 1 = 1 \cdot a = a$$

For matrix multiplication, there is, as we have seen, a neutral element if matrices are square (cf. Eq. (5.5)). Therefore, we can formally define the **inverse matrix** $\mathbf{A}^{-1}$ for an $n \times n$ matrix $\mathbf{A}$:

$$\mathbf{A} \cdot \mathbf{A}^{-1} = \mathbf{A}^{-1} \cdot \mathbf{A} = \mathbb{1}_{n \times n}. \tag{5.9}$$

Example The two matrices

$$\mathbf{A} = \begin{pmatrix} -2 & 1 \\ \frac{3}{2} & -\frac{1}{2} \end{pmatrix} \text{ and } \mathbf{A}^{-1} = \begin{pmatrix} 1 & 2 \\ 3 & 4 \end{pmatrix}$$

fulfill the condition:

$$\mathbf{A} \cdot \mathbf{A}^{-1} = \begin{pmatrix} -2 & 1 \\ \frac{3}{2} & -\frac{1}{2} \end{pmatrix} \cdot \begin{pmatrix} 1 & 2 \\ 3 & 4 \end{pmatrix} = \begin{pmatrix} 1 & 0 \\ 0 & 1 \end{pmatrix} = \mathbb{1}_{2\times 2}.$$

The questions of the existence, determination and applications of the inverse matrix require knowledge of the handling of linear equation systems. Therefore, we postpone these points to Sect. 5.3.4 and present here only calculation rules for the inverse matrix.

▶ **Calculation Rules for Inverse Matrices**

$$(\mathbf{A}^{-1})^{-1} = \mathbf{A} \tag{5.10}$$

$$(\mathbf{A}^{-1})^\top = (\mathbf{A}^\top)^{-1} \tag{5.11}$$

$$(\mathbf{A} \cdot \mathbf{B})^{-1} = \mathbf{B}^{-1} \cdot \mathbf{A}^{-1}. \tag{5.12}$$

We leave the proof of these relationships to the reader as an exercise.

5.3 Systems of Linear Equations

5.3.1 Basic Considerations

Under a **system of linear equations** we understand a system of several linear equations for the variables $x_1, x_2, \ldots, x_n$. These equations are *linear*, if all occurring terms for the variables are linear. In the equation system there may therefore be no powers of the variables greater than 1 and no products of the form $x_i \cdot x_j$. For the n variables x_i, $i = 1, 2, \ldots, n$ there may be m equations in general. We write for the system of linear equations in matrix notation:

$$\mathbf{A} \cdot \mathbf{x} = \mathbf{b},$$

where $\mathbf{A}$ is an $m \times n$-matrix, $\mathbf{x}$ a vector with n components and $\mathbf{b}$ a vector with m components. Written out, the system of linear equations looks like this:

$$a_{11}x_1 + a_{12}x_2 + \cdots + a_{1n}x_n = b_1$$
$$a_{21}x_1 + a_{22}x_2 + \cdots + a_{2n}x_n = b_2$$
$$\vdots \qquad \vdots$$
$$a_{m1}x_1 + a_{m2}x_2 + \cdots + a_{mn}x_n = b_m$$

with $n, m \in \mathbb{N}$. n is the number of unknowns, m the number of equations.

5.3 Systems of Linear Equations

Under the **solution** of a system of linear equation we understand a vector x_L, for which all equations of the system of linear equations are fulfilled, i.e. true statements are made.

The solution behavior of system of linear equations with any number of variables and equations can be divided into three categories:

1. There is a unique solution to the system, i.e. by the vector $x = x_L$ all equations become true statements.
2. The system has no solution, i.e. there is no vector x_L of the kind that all equations turn into a true statement.
3. The system is multivalued solvable, i.e. the solution vector x_L contains at least one freely selectable parameter.

An a system of linear equations with the same number of unknowns as equations ($n = m$) is called square, a system with more equations than unknowns ($m > n$) is called an **overdetermined system of linear equations**. Overdetermined equation systems usually—i.e. there are exceptions—have no solution.

A system of linear equations with fewer equations than unknowns ($m < n$) is called an **underdetermined linear system**. In general, underdetermined linear systems have infinitely many solutions, the solution vector contains one or more freely selectable parameters.[4]

The three solution categories can be examined in simple linear systems with two variables and two equations and also graphically represented for this case.

We consider the linear system:

$$-3x_1 + 2x_2 = 4 \tag{5.13}$$

$$x_2 = 5. \tag{5.14}$$

If we insert Eqs. (5.14) into (5.13), we get:

$$-3x_1 + 10 = 4$$

$$x_1 = 2.$$

The solution of the linear system is therefore:

$$x_L = \begin{pmatrix} 2 \\ 5 \end{pmatrix}.$$

In the graphical representation, you get an intersection of the two lines (5.13) and (5.14). This is shown in Fig. 5.5.

---

[4] The determination of whether a linear system is square, under- or overdetermined requires further techniques for the investigation of the equations, in particular the concept of the rank of a matrix. We refer at this point to the literature, see Arens et al. (2018) Chap. 14, or Deisenroth et al. (2020).

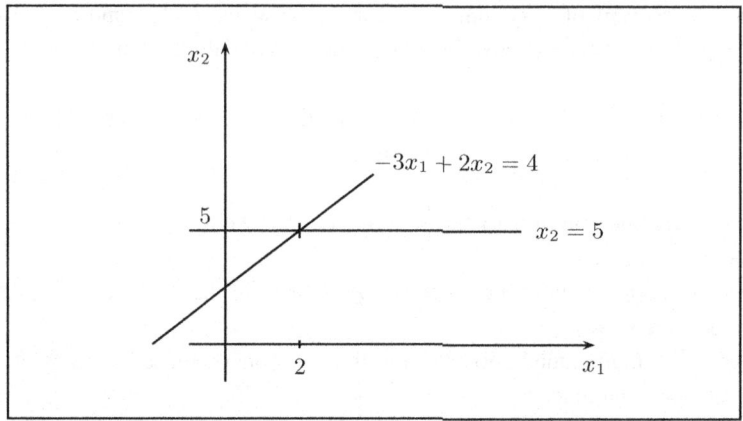

Fig. 5.5 Graphical representation of the unique solution of a linear equation system

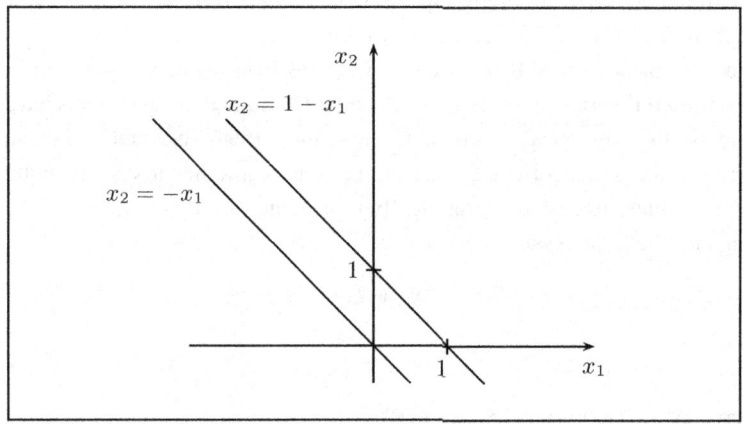

Fig. 5.6 Graphical representation of an unsolvable linear equation system

Let us now consider a second system of linear equations:

$$x_1 + x_2 = 0 \tag{5.15}$$

$$x_1 + x_2 = 1. \tag{5.16}$$

From (5.15) it follows that $x_1 = -x_2$. If we use this in Eq. (5.16), we obtain $-x_2 + x_2 = 1$, thus yielding the false statement $0 = 1$. This linear equation system is therefore unsolvable. In the graphical representation in Fig. 5.6 this situation is expressed in the fact that the two lines are parallel and do not intersect. Finally, the third case. We consider the system of linear equations:

$$x_1 + 2x_2 = 3 \tag{5.17}$$

5.3 Systems of Linear Equations

$$2x_1 + 4x_2 = 6. \tag{5.18}$$

Now let us set $x_1 = 3 - 2 x_2$ from Eq. (5.17) in Eq. (5.18) such that we obtain an always true statement, for example in the form

$$2(3 - 2x_2) + 4x_2 = 6$$

$$x_2 = x_2.$$

We write the original system in the form:

$$x_1 = 3 - 2x_2$$

$$x_2 = x_2$$

and interpret the solution as follows: For x_2 an arbitrary real parameter $x_2 = t \in \mathbb{R}$ can be chosen, for the solution x_L we get:

$$\mathbf{x}_L = \begin{pmatrix} 3 - 2t \\ t \end{pmatrix}.$$

In the graphical representation (cf. Fig. 5.7) one recognizes that the Eqs. (5.17) and (5.18) lead to identical lines, there are therefore arbitrarily many intersection points. In this case one says that the equations of the system of linear equations are *linearly dependent*. This expresses that the vectors formed from the coefficients are linearly dependent. Here these are the vectors

$$\begin{pmatrix} 1 & 2 & 3 \end{pmatrix} \quad \text{and} \quad \begin{pmatrix} 2 & 4 & 6 \end{pmatrix},$$

where the third component always stands for the right side of the Eqs. (5.17) and (5.18).

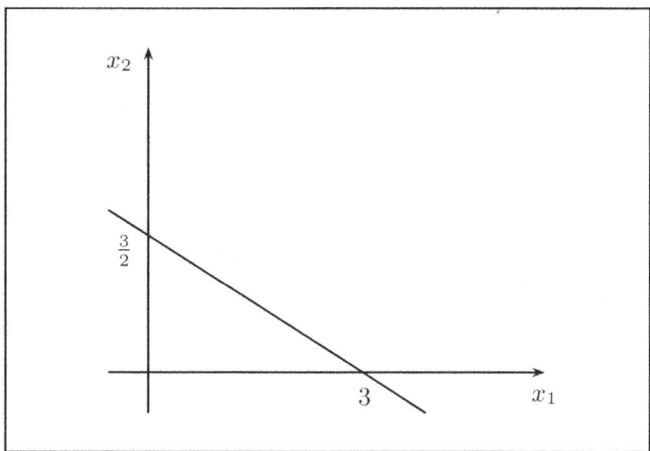

Fig. 5.7 Graphical representation of an underdetermined linear system

5.3.2 Solution Methods for Systems of Linear Equations

In this chapter we examine two methods for solving linear equation systems.

First, we consider the **Gaussian Elimination Method**. The idea is to successively eliminate variables to bring the system of linear equations into a form in which only one variable occurs in the last equation, two variables in the penultimate equation, and all variables in the first equation. We consider an example of a linear equation system with three variables and three equations:

$$2x_1 + 4x_2 - 2x_3 = 20 \tag{5.19}$$

$$x_1 + 3x_2 - 2x_3 = 13 \tag{5.20}$$

$$-x_1 + x_2 + x_3 = 5. \tag{5.21}$$

In a first step, we eliminate the variable x_1 from all equations except the first. It should be noted that the other equations are used at least once. By subtracting Eqs. (5.19)–(5.20) and adding Eqs. (5.20) and (5.21) we obtain the equivalent system of linear equations:[5]

$$\tfrac{1}{2} \cdot (5.19): \qquad x_1 + 2x_2 - x_3 = 10 \tag{5.22}$$

$$(5.19) - (5.20): \qquad -x_2 + x_3 = -3 \tag{5.23}$$

$$(5.20) + (5.21): \qquad 4x_2 - x_3 = 18. \tag{5.24}$$

In the next step, we want to achieve that from (5.24) an equation results in which only the variable x_3 occurs. Furthermore, it is desirable that the first variables in the corresponding equations have the coefficient 1. We achieve this by:

$$(5.22): \qquad x_1 + 2x_2 - x_3 = 10 \tag{5.25}$$

$$-(5.23): \qquad x_2 - x_3 = 3 \tag{5.26}$$

$$\tfrac{1}{3}(4 \cdot (5.23) + (5.24)): \qquad x_3 = 2. \tag{5.27}$$

The solution to the original system can now be determined step by step from below:

$$\text{from } (5.27): \quad x_3 = 2$$
$$\text{using this and } (5.26): \quad x_2 = 3 + x_3 = 5$$
$$\text{using this and } (5.25): \quad x_1 = 10 - 2x_2 + x_3 = 2$$

---

[5]The following transformations of the linear equation systems are equivalent transformations that we have studied in Sect. 1.4, such transformations do not change the solution set.

5.3 Systems of Linear Equations

This gives the solution vector

$$\mathbf{x}_L = \begin{pmatrix} 2 \\ 5 \\ 2 \end{pmatrix}.$$

In the Gaussian elimination method, the variables are eliminated step by step, resulting in a "triangular form" of the system of linear equations (which are Eqs. (5.25)–(5.27)), from which the solution can be determined in a few elementary steps.[6]

A second method that we will look at here is very similar to the Gaussian method, but aims at a slightly different representation of the system of linear equations in order to determine the solution. In this method, which is called the **Pivot method** or **Method of complete elimination of variables**[7], the system of linear equations is brought into the so-called **Diagonal form**. The solution of the original system of linear equations can be read directly from this. If we consider the coefficients of the system of linear equations as column vectors, then in this method unit vectors are generated step by step. To illustrate this method, we consider the same example as before to make the differences clear.

$$2x_1 + 4x_2 - 2x_3 = 20 \tag{5.28}$$

$$x_1 + 3x_2 - 2x_3 = 13 \tag{5.29}$$

$$-x_1 + x_2 + x_3 = 5. \tag{5.30}$$

First, the unit vector is generated in column 1 (coefficients of the variable x_1). In the transformation, only the equation with the 1 as coefficient and the equation with the 0 as coefficient are used in this process.

$$\frac{1}{2} \cdot (5.28): \qquad 1 \cdot x_1 + 2 \cdot x_2 - x_3 = 10 \tag{5.31}$$

$$(5.29) - (5.31): \qquad 0 \cdot x_1 + x_2 - x_3 = 3 \tag{5.32}$$

$$(5.30) + (5.31): \qquad 0 \cdot x_1 + 3 \cdot x_2 - 0 \cdot x_3 = 15. \tag{5.33}$$

In the next step, we are interested in generating the column vector

$$\mathbf{e}_2 = \begin{pmatrix} 0 \\ 1 \\ 0 \end{pmatrix},$$

---

[6] See also Arens et al. (2018), Sect. 14.2.

[7] The pivot method is particularly used in the simplex method for solving linear programs. See the literature on operations research, A. Koop and Moock (2018), W. Domschke et al. (2015) or Hillier and Lieberman (2010).

whereby of course the unit vector

$$\mathbf{e}_1 = \begin{pmatrix} 1 \\ 0 \\ 0 \end{pmatrix}$$

is to be retained. This is achieved by:

$(5.31) - 2 \cdot (5.32):$ $\qquad 1 \cdot x_1 + 0 \cdot x_2 + x_3 = 4 \qquad$ (5.34)

$(5.32):$ $\qquad 0 \cdot x_1 + 1 \cdot x_2 - x_3 = 3 \qquad$ (5.35)

$(5.33) - 3 \cdot (5.32):$ $\qquad 0 \cdot x_1 + 0 \cdot x_2 + 3 \cdot x_3 = 6. \qquad$ (5.36)

The last unit vector

$$\mathbf{e}_3 = \begin{pmatrix} 0 \\ 0 \\ 1 \end{pmatrix}$$

is obtained by the transformation:

$(5.34) - \frac{1}{3} \cdot (5.36):$ $\qquad 1 \cdot x_1 + 0 \cdot x_2 + 0 \cdot x_3 = 2 \qquad$ (5.37)

$(5.35) + \frac{1}{3} \cdot (5.36):$ $\qquad 0 \cdot x_1 + 1 \cdot x_2 + 0 \cdot x_3 = 5 \qquad$ (5.38)

$\frac{1}{3} \cdot (5.36):$ $\qquad 0 \cdot x_1 + 0 \cdot x_2 + 1 \cdot x_3 = 2. \qquad$ (5.39)

With this method, we can read the solution

$$\mathbf{x}_L = \begin{pmatrix} 2 \\ 5 \\ 2 \end{pmatrix}$$

directly.

We want to look at this process, the pivot process, again from the point of view of schematization. We write the coefficients in a table and carry out the calculation process analogously to the transformations of the system of linear equations above.

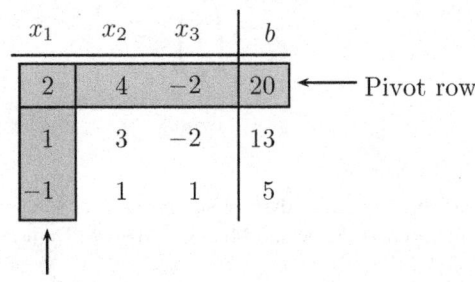

5.3 Systems of Linear Equations

The column for which the unit vector is to be generated defines the **pivot column**. The row in which the 1 is to stand defines the **pivot row**. The element that intersects the pivot column and the pivot row is called the **pivot element**. All other elements of the pivot column are referred to as **pivot column coefficients** (pcc) of a row. The transformation of the system of equations is carried out row by row in two steps:

▶ **Definition (Pivot Step)**

1. Transformation of the pivot row:

$$\text{new pivot row} = \frac{\text{old pivot row}}{\text{pivot element}}.$$

2. Transformation of all other rows:

$$\text{new row} = \text{old row} - \text{pcc} \cdot \text{new pivot row}.$$

This transformation of the table corresponds exactly to the transformations of the system of linear equations.

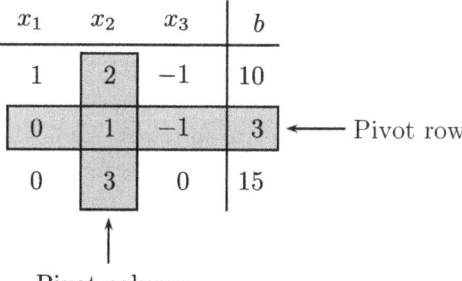

In this table, pivot column and row are defined again and another transformation step is performed:

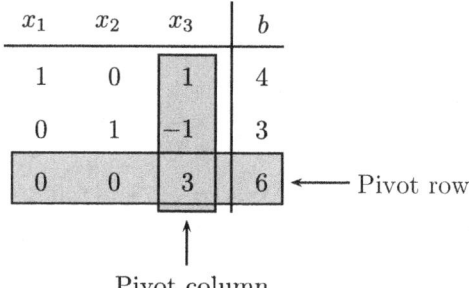

In the last step, we get the diagonal form of the system of linear equations:

x_1	x_2	x_3	b
1	0	0	2
0	1	0	5
0	0	1	2

A more general formulation of the transformation rule can be obtained by using an index form of the system of linear equations:

x_1	x_2	$\cdots$	x_k	$\cdots$	x_p	$\cdots$	x_n	b
a_{11}	a_{12}	$\cdots$	a_{1k}	$\cdots$	a_{1p}	$\cdots$	a_{1n}	b_1
a_{21}	a_{22}	$\cdots$	a_{2k}	$\cdots$	a_{2p}	$\cdots$	a_{2n}	b_2
$\vdots$	$\vdots$	$\vdots$	$\vdots$	$\vdots$	$\vdots$	$\vdots$	$\vdots$	$\vdots$
a_{i1}	a_{i2}	$\cdots$	$\boxed{a_{ik}}$	$\cdots$	a_{ip}	$\cdots$	a_{in}	b_i
$\vdots$	$\vdots$	$\vdots$	$\vdots$	$\vdots$	$\vdots$	$\vdots$	$\vdots$	$\vdots$
a_{j1}	a_{j2}	$\cdots$	a_{jk}	$\cdots$	a_{jp}	$\cdots$	a_{jn}	b_j
a_{m1}	a_{m2}	$\cdots$	a_{mk}	$\cdots$	a_{mp}	$\cdots$	a_{mn}	b_m

The k-th column is the pivot column, the i-th row of the table is the pivot row; a_{ik} is then the pivot element. For the transformation of the elements in the pivot row, the following calculation applies:

$$a_{ip}^{\text{new}} = \frac{a_{ip}}{a_{ik}}, \tag{5.40}$$

$$b_i^{\text{new}} = \frac{b_i}{a_{ik}}. \tag{5.41}$$

In order to generate a 0 in the pivot column in rows $j \neq i$, which are not pivot rows, a_{jk} must be transformed using a_{jk} and a_{ik} according to:

$$a_{jk}^{\text{new}} \stackrel{!}{=} 0 = a_{jk} - \frac{a_{jk}}{a_{ik}} \cdot a_{ik}, \quad \text{for } j \neq i.$$

A similar transformation must therefore be carried out for all row elements a_{jp} that are not in the pivot column $p \neq k$:

$$a_{jp}^{\text{new}} = a_{jp} - \frac{a_{jk}}{a_{ik}} \cdot a_{ip}, \quad p = 1, 2, \ldots, n \tag{5.42}$$

5.3 Systems of Linear Equations

$$b_j^{\text{new}} = b_j - \frac{a_{jk}}{a_{ik}} \cdot b_i, \qquad (5.43)$$

with

$$j = 1, 2, \ldots, m; \quad p = 1, 2, \ldots, n; \quad j \neq i; p \neq k.$$

We will now look at two examples to see how the cases of insoluble and non-unique system of equations are represented in the pivot method.

x_1	x_2	x_3	b
1	2	−1	10
1	3	−2	13
−1	−4	3	−12

x_1	x_2	x_3	b
1	2	−1	10
0	1	−1	3
0	−2	2	−2

x_1	x_2	x_3	b
1	0	1	4
0	1	−1	3
0	0	0	4

If we look at the last table, we see that no further pivot step is possible because the pivot element in question has the value 0. The process must be interrupted at this point. When written as an equation, it becomes clear that the last row represents a false statement:

$$0 \cdot x_1 + 0 \cdot x_2 + 0 \cdot x_3 = 4$$
$$0 = 4.$$

Thus the system is insoluble.

In a corresponding way, the case of non-uniquely solvable system of linear equations does not present itself clearly, as the following example shows:

x_1	x_2	x_3	b
1	2	−1	10
1	3	−2	13
−1	−4	3	−16

x_1	x_2	x_3	b
1	2	−1	10
0	1	−1	3
0	−2	2	−6

x_1	x_2	x_3	b
1	0	1	4
0	1	−1	3
0	0	0	0

Again, no further step is possible here, but the last line contains a true statement. If we write $0 = x_3 - x_3$ for this, we get the general solution:

$$x_1 + x_3 = 4$$
$$x_2 - x_3 = 3$$
$$0 = x_3 - x_3$$

or:

$$\begin{pmatrix} x_1 \\ x_2 \\ x_3 \end{pmatrix} = \begin{pmatrix} 4 - x_3 \\ 3 + x_3 \\ x_3 \end{pmatrix}$$

with $x_3 \in \mathbb{R}$ as any parameter.

5.3.3 Standardized form of Systems of Linear Equations

We consider the general Systems of Linear Equations

$$A \cdot x = b$$

with:

$$x = \begin{pmatrix} x_1 \\ x_2 \\ \vdots \\ x_n \end{pmatrix} \quad \text{and} \quad b = \begin{pmatrix} b_1 \\ b_2 \\ \vdots \\ b_m \end{pmatrix}.$$

We can introduce unit vectors in a pivot process until the table takes the following form after k steps:[8]

x_1	x_2	$\cdots$	x_k	x_{k+1}	$\cdots$	x_n	b
1	0	$\cdots$	0				$\tilde{b}_1$
0	1	$\cdots$	0				$\tilde{b}_2$
$\vdots$	$\vdots$	$\vdots$	$\vdots$		Ω		$\vdots$
0	0	$\cdots$	1				$\tilde{b}_k$
0	0	$\cdots$	0	0	$\cdots$	0	$\tilde{b}_{k+1}$
$\vdots$	$\vdots$	$\vdots$	$\vdots$	$\vdots$		$\vdots$	$\vdots$
0	0	$\cdots$	0	0	$\cdots$	0	$\tilde{b}_m$

The left part of the table $(x_1, x_2, \ldots, x_k)$ contains the unit vectors generated in the process. In the field bottom right, all elements must be zero, otherwise another pivot step would be possible. In the table designated with Ω, there are any coefficients that arise during the transformation.

The solution behavior of a linear equation system can now be read off from the table:

1. The system is unsolvable if one of the values $\tilde{b}_{k+1}, \tilde{b}_{k+2}, \ldots, \tilde{b}_m$ is different from zero.
2. The system is uniquely solvable if $k = m = n$ is.
3. For $\tilde{b}_{k+1} = \tilde{b}_{k+2} = \ldots = \tilde{b}_m = 0$ the system of linear equations is ambigously solvable, $(m - k)$ free solution parameters can be introduced.

In the two examples considered above, $k = 2$ each time. Once $\tilde{b}_3 = 4$ resulted. The system of linear equations is therefore unsolvable, in the last case $\tilde{b}_3 = 0$, and thus a solution set that has a free parameter.

---

[8] Under certain circumstances, this form only results after exchanging rows, which, however, does not change the solution of the LGS, since exchanging the rows only changes the order of the equations.

5.3.4 Matrix Inversion

The pivot method for solving linear equation systems, which was treated in the previous section, provides an efficient way to determine the inverse $\mathbf{A}^{-1}$ of a matrix $\mathbf{A}$. We recall the definition of the inverse matrix:

$$\mathbf{A} \cdot \mathbf{A}^{-1} = \mathbb{1}_{n \times n}. \tag{5.44}$$

We can interpret matrix multiplication as the scalar product of the row vectors of $\mathbf{A}$ with the column vectors of $\mathbf{A}^{-1}$. The first column vector of the identity matrix therefore results from the product of the matrix $\mathbf{A}$ with the first column vector of $\mathbf{A}^{-1}$.

This is considered in more detail below. The matrix $\mathbf{A}$ is given by:

$$\mathbf{A} = \begin{pmatrix} a_{11} & a_{12} & \cdots & a_{1n} \\ a_{21} & a_{22} & \cdots & a_{2n} \\ \vdots & \vdots & \cdots & \vdots \\ a_{n1} & a_{n2} & \cdots & a_{nn} \end{pmatrix}.$$

For the coefficients of the matrix $\mathbf{A}^{-1}$ to be determined, we introduce the unknowns x_{ij}:

$$\mathbf{A}^{-1} = \begin{pmatrix} x_{11} & x_{12} & \cdots & x_{1n} \\ x_{21} & x_{22} & \cdots & x_{2n} \\ \vdots & \vdots & \cdots & \vdots \\ x_{n1} & x_{n2} & \cdots & x_{nn} \end{pmatrix} = \begin{pmatrix} \mathbf{x}_1 & \mathbf{x}_2 & \cdots & \mathbf{x}_n \end{pmatrix}.$$

We write the identity matrix in the form:

$$\mathbb{1}_{n \times n} = \begin{pmatrix} 1 & 0 & \cdots & 0 \\ 0 & 1 & \cdots & 0 \\ \vdots & \vdots & \vdots & \vdots \\ 0 & 0 & \cdots & 1 \end{pmatrix} = \begin{pmatrix} \mathbf{e}_1 & \mathbf{e}_2 & \cdots & \mathbf{e}_n \end{pmatrix}.$$

With these transformations, Eq. (5.44) can be separated into n systems of linear equations. For each column vector $\mathbf{x}_i$ of $\mathbf{A}^{-1}$, a systems of linear equations of the following form has to be solved:

$$\mathbf{A} \cdot \mathbf{x}_i = \mathbf{e}_i, \quad i = 1, 2, \ldots, n.$$

These systems all have the same coefficient matrix $\mathbf{A}$ and only differ in the terms on the right-hand side. The pivot method therefore runs for all n systems in a similar way, only the right-hand side of the table is different.

5.3 Systems of Linear Equations

Example We calculate the inverse matrix of:

$$A = \begin{pmatrix} 1 & 2 \\ 2 & 1 \end{pmatrix}.$$

It is to be solved:

$$\begin{pmatrix} 1 & 2 \\ 2 & 1 \end{pmatrix} \cdot \begin{pmatrix} x_{11} & x_{12} \\ x_{21} & x_{22} \end{pmatrix} = \begin{pmatrix} 1 & 0 \\ 0 & 1 \end{pmatrix}.$$

This equation corresponds to the two systems of linear equations:

$$x_{11} + 2x_{21} = 1$$
$$2x_{11} + x_{21} = 0$$

and

$$x_{12} + 2x_{22} = 0$$
$$2x_{12} + x_{22} = 1.$$

We solve the first one in the form:

x_{11}	x_{21}	b		x_{11}	x_{21}	b		x_{11}	x_{21}	b
1	2	1		1	2	1		1	0	$-\frac{1}{3}$
2	1	0		0	−3	−2		0	1	$\frac{2}{3}$

The second one accordingly:

x_{12}	x_{22}	b		x_{12}	x_{22}	b		x_{12}	x_{22}	b
1	2	0		1	2	0		1	0	$\frac{2}{3}$
2	1	1		0	−3	1		0	1	$-\frac{1}{3}$

The coefficients in the left part of the tables are thus transformed according to the pivot method (even if the corresponding variables are different). Only the right columns are different in the solution. We now summarize the solution of the two systems of linear equations in a table and write the matrix **A** on the left, on the right the unit matrix results.

A		E								E		A^{-1}	
1	2	1	0		1	2	1	0		1	0	$-\frac{1}{3}$	$\frac{2}{3}$
2	1	0	1		0	−3	−2	1		0	1	$\frac{2}{3}$	$-\frac{1}{3}$

The framed element ☐ denotes the pivot element in the respective step. In the last step, the inverse of the matrix **A** is obtained. The existence of the inverse $\mathbf{A}^{-1}$ arises during the pivot process. The pivot process terminates if, during the course of the process, no pivot element $\neq 0$ exists anymore. Then the matrix is not invertible. To this end, we consider the following example:

A			E		
☐1	2	2	1	0	0
2	1	1	0	1	0
1	1	1	0	0	1

$\longrightarrow$

1	2	2	1	0	0
0	☐-3	-3	-2	1	0
0	-1	-1	-1	0	-1

$\longrightarrow$

1	0	0	$-\frac{2}{3}$	$\frac{2}{3}$	0
0	1	1	$\frac{2}{3}$	$-\frac{1}{3}$	0
0	0	0	$-\frac{1}{3}$	$-\frac{1}{3}$	0

It is not possible to generate the third required unit vector. Of course, this does not depend on the order in which the unit vectors are generated in the pivot process. The reader can easily convince himself that the process terminates in the same way if the unit vector $\mathbf{e}_3 = \begin{pmatrix} 0 \\ 0 \\ 1 \end{pmatrix}$ is generated first.

The existence of an inverse matrix is of interest in connection with systems of linear equations. Let us consider the equation system:

$$\mathbf{A} \cdot \mathbf{x} = \mathbf{b}$$

and multiply this equation by $\mathbf{A}^{-1}$ from the left, we obtain:

$$\underbrace{\mathbf{A}^{-1} \cdot \mathbf{A}}_{\mathbf{1}_{n \times n}} \mathbf{x} = \mathbf{A}^{-1} \mathbf{b} \tag{5.45}$$

$$\mathbf{x} = \mathbf{A}^{-1} \mathbf{b}.$$

The last transformation requires that $n = m$, that is, the number of equations is equal to the number of unknowns. It shows that the existence of a matrix $\mathbf{A}^{-1}$ is associated with

5.3 Systems of Linear Equations

the unique solvability of the equation system. Neither in the solution of the equation system nor in the inversion of the matrix **A** can a complete diagonalization be carried out.

There is a characteristic for the existence of an inverse matrix, which we want to consider in conclusion: The **determinant** of a matrix.

▶ **Definition (Determinant of a 2×2 Matrix)**

The determinant of a square 2×2 matrix is defined by

$$\det \mathbf{A} = \det \begin{pmatrix} a_{11} & a_{12} \\ a_{21} & a_{22} \end{pmatrix} = a_{11}a_{22} - a_{12}a_{21} \in \mathbb{R}. \tag{5.46}$$

The determinant of a matrix is therefore a real number. In the case of inverting the matrix **A** it turns out that the inverse exists only if $\det \mathbf{A} \neq 0$. The determinant of the coefficient matrix of a system of linear equations can also be used to decide whether there is a unique solution of the system, as stated here without proof:

A linear equation system of the form $\mathbf{Ax} = \mathbf{b}$ is uniquely solvable if $\det \mathbf{A} \neq 0$. This statement can be generalized to $n \times n$ matrices, as indicated here without proof.[9] With the determinant of a matrix **A** the existence of the inverse matrix can be easily checked. For 3×3 matrices, the determinant can be calculated according to the following scheme:

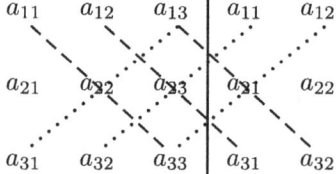

We write the first two columns of the matrix to the right of the columns of the matrix and then take the difference of the sum of products that are on the corresponding diagonals.

$$\det \mathbf{A} = (a_{11}a_{22}a_{33} + a_{12}a_{23}a_{31} + a_{13}a_{21}a_{32}) \\ - (a_{13}a_{22}a_{31} + a_{11}a_{23}a_{32} + a_{12}a_{21}a_{33}). \tag{5.47}$$

Determinants of $n \times n$ matrices are reduced to determinants of $(n-1) \times (n-1)$ matrices according to the Laplace expansion theorem and can thus be calculated iteratively.[10]

---

[9] See, for example, Goebbels and Ritter (2011), Sect. 1.8.
[10] See Bronstein (2005) or Goebbels and Ritter (2011), Sect. 1.8.

5.3.5 Business Applications

▶ **Production Interdependence**

In complex production processes that occur, for example, in the chemical industry, the problem arises that intermediate products are brought to the market on the one hand, and on the other hand, they are also used as a product to be further processed within the production process. Such a intertwining of products over several production stages is often represented in the so-called **Gozinto graph**[11]. We consider two production nodes i and j from which the quantities x_i and x_j are produced. The size a_i specifies how much of the product i is delivered. The coefficient k_{ij} specifies how many units of the product i are required for the production of one unit of the product j.

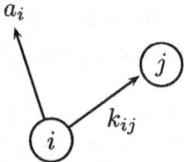

For the calculation of x_i therefore applies:

$$x_i = a_i + \sum_j k_{ij} \cdot x_j.$$

Given deliveries a_i and entanglement coefficients k_{ij} the vector of production quantities $\mathbf{x}^\top = (x_1, x_2, \ldots, x_n)$ results in a linear equation system. We consider this with an example that is described by the following Gozinto graph:

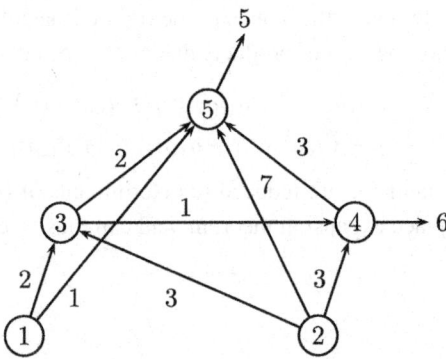

---

[11] According to A. Vazsonyi, who, in a humorous way, attributed these considerations to the "Italian mathematician" *Zepartzat Gozinto*. Pronounced *The part that goes into* makes the connection clear.

5.3 Systems of Linear Equations

This entanglement is described by the following equation system:

$$x_1 = 2x_3 + x_5$$
$$x_2 = 3x_3 + 3x_4 + 7x_5$$
$$x_3 = x_4 + 2x_5$$
$$x_4 = 3x_5 + 6$$
$$x_5 = 5.$$

Due to the special form of the equation system, the solution is easy to read here:

$$x_5 = 5$$
$$x_4 = 3 \cdot 5 + 6 = 21$$
$$x_3 = 21 + 2 \cdot 5 = 31$$
$$x_2 = 3 \cdot 31 + 3 \cdot 21 + 7 \cdot 5 = 191$$
$$x_1 = 2 \cdot 31 + 5 = 67.$$

▶ **Technology Matrix and Leontief Inverse** We generalize the above consideration to a intertwining of different producing sectors, as they are for example also given in the national economy.[12] The production coefficients a_{ij} specify how many units of the i-th product are needed for the production of one unit of the j-th product. This is called the **endogenous input**. For the calculation of the total production of all $i = 1, 2, \ldots, n$ products, the final consumption b_i of the individual products is added.

For the total production one obtains in matrix notation:

$$\mathbf{x} = \mathbf{A} \cdot \mathbf{x} + \mathbf{b}.$$

If the production $\mathbf{x}$ is fixed, the final consumption can be calculated according to

$$\mathbf{b} = (\mathbf{E} - \mathbf{A}) \cdot \mathbf{x}. \tag{5.48}$$

The matrix $\mathbf{E} - \mathbf{A}$ is called the **technology matrix**.

If one assumes, however, that final consumption $\mathbf{b}$ is given, then production is calculated according to:

$$\mathbf{x} = (\mathbf{E} - \mathbf{A})^{-1} \cdot \mathbf{b}. \tag{5.49}$$

The matrix $(\mathbf{E} - \mathbf{A})^{-1}$ is called the **Leontief inverse**. From this context it can be seen that, for any given final consumption $\mathbf{b}$, production $\mathbf{x}$ can only be calculated if the technology matrix is invertible (i.e. the Leontief inverse exists) and all elements of $(\mathbf{E} - \mathbf{A})^{-1}$ are positive or zero.

---

[12] Wassily Leontief (1905–1999), Nobel Prize for Economics 1973.

▶ **Internal Performance Accounting**

Operational performance is delivered in order to create products or services. For this purpose, it is usually necessary that some departments also provide services to each other. Examples include personnel services or energy provision. For the determination of adequate remuneration prices, the **internal performance accounting** is carried out. The value of the produced performance consists of the **primary costs**, which arise for the performance and the **secondary costs**, which result from the services received. For price calculation, remuneration prices are introduced, which at the same time enter into the secondary costs and the value of the service. The relationship is therefore:

$$\text{primary costs} + \text{secondary cost} = \text{value of the produced performance},$$

with

$$\text{secondary costs} = \text{received performance} \times \text{transfer price}$$

and

$$\text{value of the produced performance} = \text{overall performance} \times \text{transfer price}$$

Therefore, a system of linear equations is required for the calculation of the remuneration prices.

Example We consider three cost centers that provide and receive services to and from each other, export services and incur primary costs.

Delivery receipt ↓ ⟶	K_1	K_2	K_3	Export	Total output	Primary Costs
K_1	0	10	0	20	30	100
K_2	8	0	0	40	50	150
K_3	4	5	20	20	40	200

For the calculation of prices, the following system results from the consideration of the costs and deliveries for each cost center:

$$K_1: \quad 100 + 0 \cdot p_1 + 8 \cdot p_2 + 4 \cdot p_3 = 30 \cdot p_1$$
$$K_2: \quad 150 + 10 \cdot p_1 + 0 \cdot p_2 + 5 \cdot p_3 = 50 \cdot p_2$$
$$K_3: \quad 200 + 0 \cdot p_1 + 0 \cdot p_2 + 20 \cdot p_3 = 40 \cdot p_3.$$

The solution results in:

$$p_3 = 10; \quad p_2 = \frac{740}{142} \approx 5{,}21; \quad p_1 \approx 6{,}05.$$

5.3.6 Eigenvalues of a Matrix

The following consideration leads to another important concept for the application of matrices. The multiplication of a matrix with a vector transforms this vector:

5.3 Systems of Linear Equations

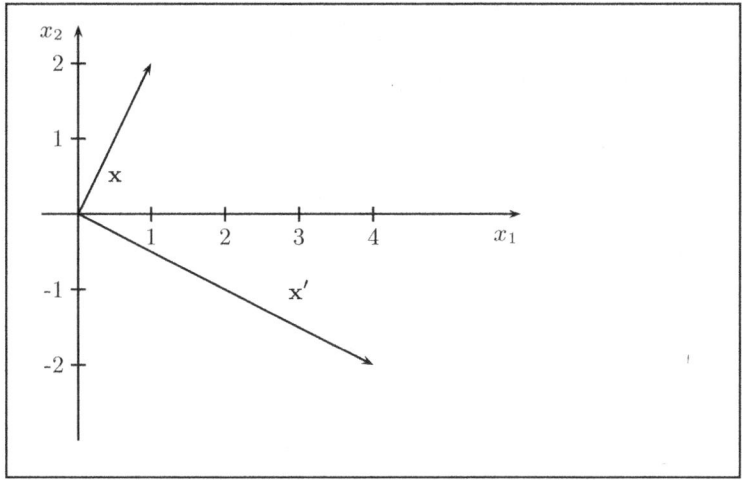

Fig. 5.8 Matrix multiplication as transformation of a vector

$$\mathbf{A} \cdot \mathbf{x} = \mathbf{x}'.$$

The Fig. 5.8 shows such a transformation of a vector for the example:

$$\begin{pmatrix} 2 & 1 \\ 0 & -1 \end{pmatrix} \cdot \begin{pmatrix} 1 \\ 2 \end{pmatrix} = \begin{pmatrix} 4 \\ -2 \end{pmatrix}.$$

We are interested in special vectors λ x_e for a given Matrix A. These special vectors have the following property: If we multiply the matrix A with the vectors x_e, the product results in a multiple $\lambda\_e \in$ R of x_e. We call these vectors eigenvectors of the matrix A and the factors λ_e eigenvalues of the matrix A. Hence, the eigenvectors have to satisfy the following matrix equation:

$$\mathbf{A} \cdot \mathbf{x}_e = \lambda_e \mathbf{x}_e \tag{5.50}$$

or

$$\left(\mathbf{A} - \lambda_e \cdot \mathbb{1}_{n \times n}\right) \cdot \mathbf{x}_e = \mathbf{0}. \tag{5.51}$$

The **eigenvalue equation** (5.50) represents a system of linear equations, for which we can immediately give a so-called trivial solution: $\mathbf{x}_e = \mathbf{0}$. There can only be further solutions if the system is not uniquely solvable. A criterion for the existence of multiple solutions we have learned in Sect. 5.3.4. The determinante of the coefficient matrix must be equal to zero. The equation for the determination of the eigenvalues therefore reads:

▶ **Definition (Eigenvalues of a matrix)**

λ_e: Eigenvalues of a matrix A are solutions of the equation

$$\det\left(\mathbf{A} - \lambda_e \cdot \mathbb{1}_{n \times n}\right) = 0. \tag{5.52}$$

Eq. (5.52) is also called the **characteristic equation**.

For this we consider the above example. The eigenvalues of the matrix

$$\mathbf{A} = \begin{pmatrix} 2 & 1 \\ 0 & -1 \end{pmatrix}$$

lead to the characteristic equation:

$$\det \begin{pmatrix} 2 - \lambda_e & 1 \\ 0 & -1 - \lambda_e \end{pmatrix} = 0$$

or

$$(2 - \lambda_e)(-1 - \lambda_e) - 0 = 0$$

with the two solutions:

$$\lambda_{e1} = 2, \qquad \lambda_{e2} = -1.$$

With the eigenvalues, the eigenvectors can then be determined from Eq. (5.50).

In the context of this book, eigenvalues are used in the formulation of sufficient conditions for the existence of local extrema for functions with several variables, eigenvectors are not further discussed.[13]

5.4 Exercises

Short solutions to the following exercises can be found in the appendix.

5.1 In a production plant, two end products are manufactured from three intermediate products. Two raw materials are used to produce the intermediate products.

The respective quantity requirements are described by the following production coefficients:

	ZP1	ZP2	ZP3
RS1	1	2	2
RS2	1	2	1

	EP1	EP2
ZP1	2	2
ZP2	1	1
ZP3	1	2

---

[13] See also Arens et al. (2018), Chap. 18, or Goebbels and Ritter (2011), Sect. 3.6.

5.4 Exercises

124 units of RS1 and 98 units of RS2 are used for production. How many units of the end products can be manufactured with complete raw material consumption?

5.2 Given are the two matrices

$$A = \begin{pmatrix} 1 & 2 & 3 \\ 0 & 3 & -1 \\ -4 & 1 & 0 \end{pmatrix} \quad B = \begin{pmatrix} 2 & 2 & 1 \\ -1 & 0 & 4 \end{pmatrix}.$$

Calculate:

(a) $A \cdot B$
(b) $B \cdot A$
(c) $A^T \cdot B$
(d) $A \cdot B^T$
(e) A^2
(f) B^2
(g) $(B^T)^2$.

5.3 Multiply the matrices if possible:

(a) $A = \begin{pmatrix} 1 & 5 & 7 \\ 3 & -2 & 1 \\ 0 & 2 & 6 \end{pmatrix}, \quad B = \begin{pmatrix} 1 & 0 \\ 2 & 2 \\ -1 & 5 \end{pmatrix}.$

(b) $A = \begin{pmatrix} 2 & 3 & 7 \\ 4 & 1 & -8 \end{pmatrix}, \quad B = \begin{pmatrix} 1 & 3 & -2 \\ 2 & 7 & 0 \end{pmatrix}.$

(c) $A = \begin{pmatrix} 2 & 0 \\ 0 & 1 \end{pmatrix}, \quad B = \begin{pmatrix} -2 & 1 \\ 0 & -1 \end{pmatrix}.$

(d) $A = \begin{pmatrix} 1 & 5 & 7 \\ 3 & -2 & 1 \end{pmatrix}, \quad B = \begin{pmatrix} 2 & 1 \\ 6 & -1 \\ 8 & 0 \end{pmatrix}.$

5.4 Simplify the following matrix products:

(a) $B^T \cdot (A \cdot B^T)^{-1}$.
(b) $A^T \cdot (A \cdot B)^T \cdot (A^{-1})^T$.

5.5 Show the calculation rules for inverse matrices, Eqs. (5.10) to (5.12).

5.6 Invert the following 2×2 matrix A. Under which condition does the inverse exist?

$$A = \begin{pmatrix} a & b \\ c & d \end{pmatrix}.$$

5.7 Examine whether the following system of equations is solvable. If so, what is the solution?

$$\begin{pmatrix} 2 & 1 & -2 & 3 \\ 3 & 2 & -1 & 2 \\ 3 & 3 & 3 & -3 \end{pmatrix} \cdot \begin{pmatrix} x_1 \\ x_2 \\ x_3 \\ x_4 \end{pmatrix} = \begin{pmatrix} 1 \\ 4 \\ 5 \end{pmatrix}.$$

5.8 How must the parameters a, b, c be chosen so that the following system of linear equations is solvable?

$$\begin{pmatrix} 1 & 2 & -3 \\ 2 & 6 & -11 \\ 1 & -2 & 7 \end{pmatrix} \cdot \begin{pmatrix} x_1 \\ x_2 \\ x_3 \end{pmatrix} = \begin{pmatrix} a \\ b \\ c \end{pmatrix}.$$

Can there be a unique solution?

5.9 Invert the following matrices if the inverses exist:

(a) $\begin{pmatrix} 1 & 2 & -1 \\ 2 & 4 & 2 \\ 1 & 1 & 1 \end{pmatrix}$.

(b) $\begin{pmatrix} 1 & 2 & -1 \\ 2 & 4 & 2 \\ -4 & -8 & -8 \end{pmatrix}$.

5.10 Determine the eigenvalues of the matrix

$$\mathbf{A} = \begin{pmatrix} 2 & 4 & -2 \\ 4 & 2 & -2 \\ -2 & -2 & -1 \end{pmatrix}.$$

5.11 The determinant has the property:

$$\det(\mathbf{A} \cdot \mathbf{B}) = \det(\mathbf{A}) \cdot \det(\mathbf{B}).$$

Show this property of the determinant explicitly for any 2×2 matrices $\mathbf{A}$, $\mathbf{B}$.

5.12 A manufacturer of multivitamin preparations advertises that its preparation contains the optimal amount of vitamins A, B and C with the units of measurement $A = 16$, $B = 20$ and $C = 18$ units. Can this also be ensured by a corresponding mixture of fruit, if the units of the vitamins in the fruit are given as follows:

	A	B	C
Apple	2	2	2
Banana	3	2	1
Pineapple	1	4	3

5.4 Exercises

If so, in what proportions must the fruit be mixed?

5.13 The price matrix **P** contains the prices of three products from different suppliers.

	P_1	P_2	P_3
L_1	5	5	3
L_2	4	4	4
L_3	3	5	4

The demand matrix **B** indicates how many units of the three products are purchased in different plants.

	W_1	W_2	W_3	W_4
P_1	100	20	20	200
P_2	70	50	40	100
P_3	100	100	40	100

Calculate the matrix product $\mathbf{K} = \mathbf{P} \cdot \mathbf{B}$ and interpret the coefficients of the product matrix.

5.14 In a company, the production factors are interwoven with each other via the following production coefficients:

	p_1	p_2	p_3
p_1	0,5	0,1	0,1
p_2	0,2	0,5	0
p_3	0,1	0,1	0,2

Each day, the quantities $x_1 = 100$, $x_2 = 200$, $x_3 = 300$ of the products p_1, p_2, p_3 are produced. How many units of product can be delivered to the market?

5.15 Given the production coefficients of a sectorally interwoven production by the matrix:

$$\mathbf{A} = \begin{pmatrix} a & 0,5 \\ 0,2 & 0,5 \end{pmatrix} \quad (a > 0).$$

(a) What condition must a fulfill so that any demand on the market can also be satisfied?
(b) Calculate the quantity to be produced for the demand $b_1 = 20$, $b_2 = 30$ at a value of $a = 2/5$.

5.16 Five production sectors $P_1, P_2, \ldots, P_5$ are interwoven with each other according to the Gozinto graph below. In addition, individual products are delivered externally as indicated.

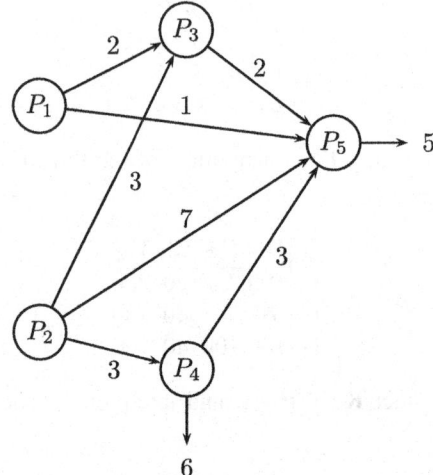

Set up a system of linear equations for the quantities to be produced and solve the system.

5.17 Which of the following products from the matrices **A** and **B**, as well as their transposed matrices, exist?

$$\mathbf{A} \cdot \mathbf{B}, \mathbf{A}^T \cdot \mathbf{B}, \mathbf{A} \cdot \mathbf{B}^T, \mathbf{A}^T \cdot \mathbf{B}^T, \mathbf{B} \cdot \mathbf{A}, \mathbf{B}^T \cdot \mathbf{A}, \mathbf{B} \cdot \mathbf{A}^T, \mathbf{B}^T \cdot \mathbf{A}^T.$$

The matrices are:

$$\mathbf{A} = \begin{pmatrix} 2 & -2 \\ 0 & 3 \end{pmatrix}, \quad \text{and} \quad \mathbf{B} = \begin{pmatrix} 1 & -1 & 0 \\ 0 & 2 & -1 \end{pmatrix}.$$

References

Arens T., Hettlich F., Karpfinger Ch., Kockelkorn U., Lichtenegger K., Stachel H. (2018): Mathematik, 4. Auflage, Spektrum Akademischer Verlag, Heidelberg.
Bronstein I.N. *et al.* (2005): Taschenbuch der Mathematik, 6. Edition, Verlag Harri Deutsch.
Deisenroth M.P., Faisal A.A., Ong C.S. (2020): Mathematics for Machine Learning, Cambridge University Press, Cambridge (UK).
Domschke W., Drexl A., Klein R., Scholl A. (2015): Einführung in Operations Research, 9. Auflage, Springer, Berlin.
Fischer G., Springborn B. (2020): Lineare Algebra, 19., vollst. überarb. u. erg. Auflage, Springer Spektrum.

References

Hillier F.S., Lieberman G.J. (2010): Introduction to Operations Research, Ninth Edition, MacGraw Hill, New York.

Koop A., Moock, H. (2018): Lineare Optimierung, Eine anwendungsorientierte Einführung in Operations Research, 2. Auflage, Spektrum Akademischer Verlag, Heidelberg.

Lang S. (1987): Linear Algebra, Third Edition, UTM, Springer Verlag, New York.

Functions with Several Variables 6

> **Learning Objectives (This Chapter Provides)**
>
> - the meaning of functions with several variables
> - the extension of differential calculus to functions with several variables
> - how to deal with extreme values with and without boundary conditions
> - the application of the gradient method
> - an introduction to the Lagrange method ◀

6.1 Introduction and Representation

In Chap. 2 we introduced the connection between an independent variable x and a dependent variable y in the form $y = f(x)$. In many cases, however, the quantity y does not only depend on one variable x, but on several variables $x_1, x_2, \ldots, x_n$.[1] We write this as:

$$y = f(\mathbf{x}) \quad \text{where} \quad \mathbf{x} = \begin{pmatrix} x_1 \\ x_2 \\ \vdots \\ x_n \end{pmatrix}.$$

This dependency is called a **function** if the uniqueness described in Chap. 2 is satisfied.

---

[1] A comprehensive representation of functions of several variables and their properties can be found, inter alia, in Arens et al. (2018) Chap. 24, Dyke (2018) or Marsden and Weinstein, Volume III (1985).

© The Author(s), under exclusive license to Springer-Verlag GmbH, DE, part of Springer Nature 2023
T. Holey and A. Wiedemann, *Analysis and Linear Algebra*,
https://doi.org/10.1007/978-3-662-66247-2_6

Instead of

$$f : \mathbb{R} \longrightarrow \mathbb{R},$$
$$x \longmapsto y = f(x)$$

we now consider the generalization:

$$f : \mathbb{R}^n \longrightarrow \mathbb{R},$$

$$\mathbf{x} \longmapsto y = f(\mathbf{x}) \quad \text{where} \quad \mathbf{x} = \begin{pmatrix} x_1 \\ x_2 \\ \vdots \\ x_n \end{pmatrix}.$$

For the case $n = 2$ the mapping results in:

$$f : \mathbb{R}^2 = \mathbb{R} \times \mathbb{R} \longrightarrow \mathbb{R},$$
$$(x_1, x_2) \longmapsto y = f(x_1, x_2).$$

In this case, the functional relationship can be graphically illustrated by the three coordinate axes x_1, x_2 and y. The function $y = f(x_1, x_2)$ represents a curved surface in space, see Fig. 6.1.

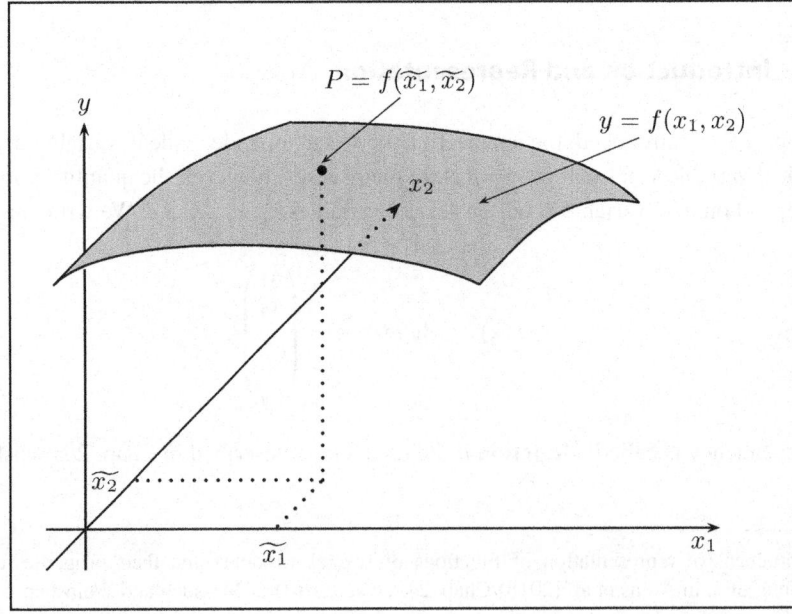

Fig. 6.1 Graphical representation of a function with two variables $f(x_1, x_2)$ as a surface in a three-dimensional space with coordinates x_1, x_2, y

6.1 Introduction and Representation

We consider the example

$$y = f(x_1, x_2) = 1 - x_1^2 - x_2^2$$

with the graphical representation in Fig. 6.2.

Another way of representing such functions arises from the consideration of cutting planes parallel to the coordinate planes. If we set $x_1 = const.$, we obtain a image of the function for each plane. In the projection onto the $y-x_2$-plane, with $x_1 = c_1 = const.$ we obtain the following representation (cf. Fig. 6.3):

$$y = 1 - c_1^2 - x_2^2.$$

We find an analogous representation in the projection onto the $y-x_1$-plane.

If we set $y = const.$, a representation of the curved surface in space results, which is known from topographical maps: The course of the profile of y over the $x_1 - x_2$-plane is described by **contour lines**.

If we set $y = c_2 = const$ in our example, then the following applies:

$$c_2 = 1 - x_1^2 - x_2^2 \quad \text{or} \quad x_1^2 + x_2^2 = 1 - c_2.$$

In this example, the contour lines are therefore concentric circles, see Fig. 6.4.

In the formulation of economic relationships, functions with several variables occur frequently. The **production function** describes the production quantity — that is, the output—as a function of production factors r_i used, such as labor, energy and raw materials.

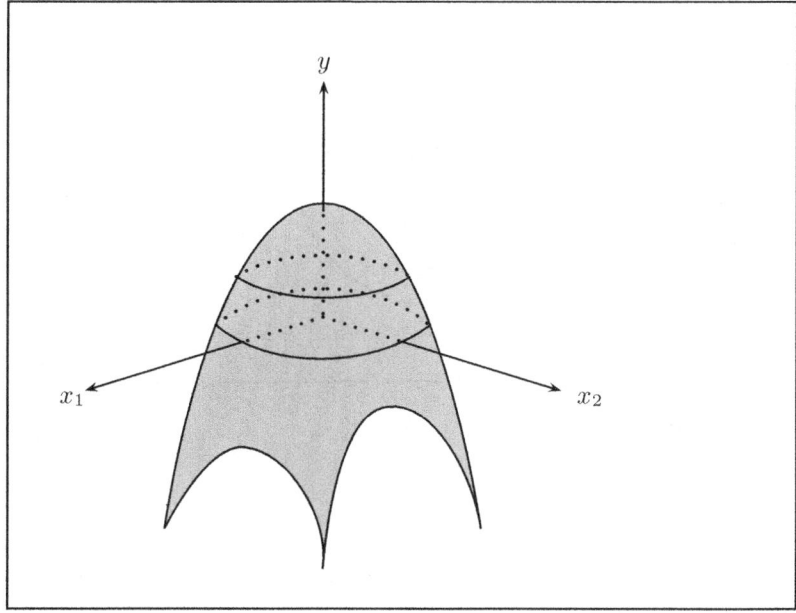

Fig. 6.2 Representation of the function $y = f(x_1, x_2) = 1 - x_1^2 - x_2^2$

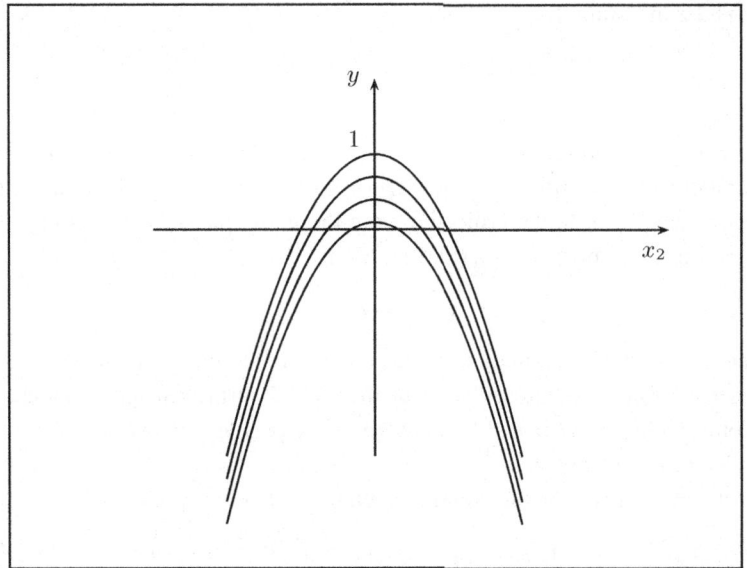

Fig. 6.3 Projection representation of the function $y = 1 - x_1^2 - x_2^2$ on the y-x_2-plane

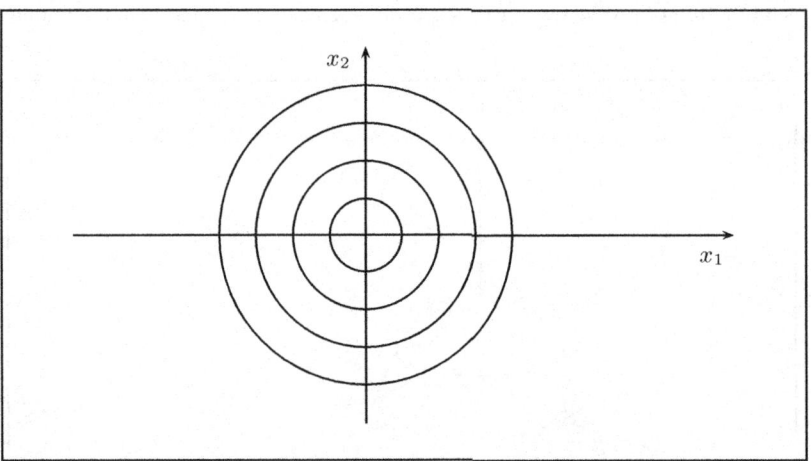

Fig. 6.4 Contour lines of the function $y = 1 - x_1^2 - x_2^2$

$$x = x(\mathbf{r})$$

Let us consider here the contour lines, that is, the lines of constant production quantities. These are called **isoquants**.

6.1 Introduction and Representation

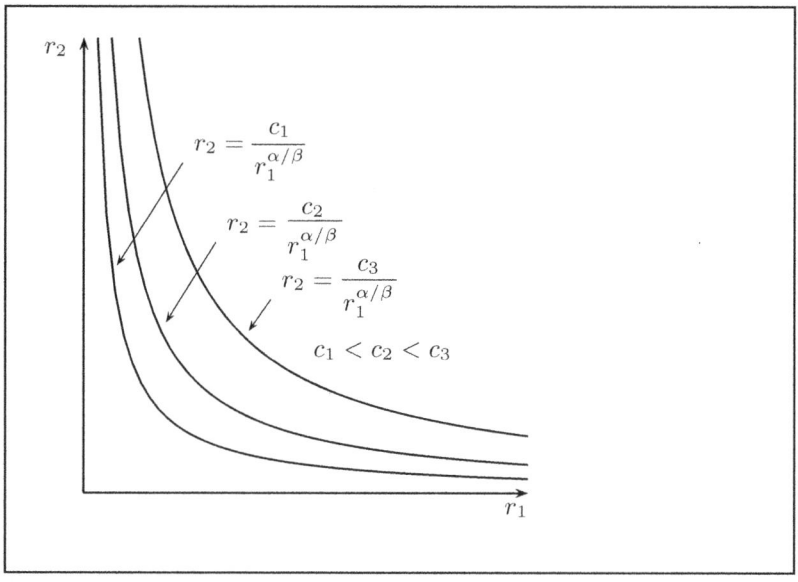

Fig. 6.5 Isoquants of the Cobb-Douglas production function

Example A production function is given by:
$$x(\mathbf{r}) = a \cdot r_1^\alpha \cdot r_2^\beta, \quad \alpha, \beta > 0.$$
The isoquants are given by the condition $x(\mathbf{r}) = const. = c$, that is,
$$c = a \cdot r_1^\alpha \cdot r_2^\beta,$$
or
$$r_2 = \left(\frac{c}{a \cdot r_1^\alpha}\right)^{1/\beta}.$$

In Fig. 6.5 three isoquants of this production function are shown with three constants c_1, c_2 and c_3.

Note:
Production functions of this form are referred to as **Cobb-Douglas production functions**.[2]

---

[2] These functions are named after the two US economists Charles Wiggins Cobb (1875-1949) and Paul Howard Douglas (1892–1976).

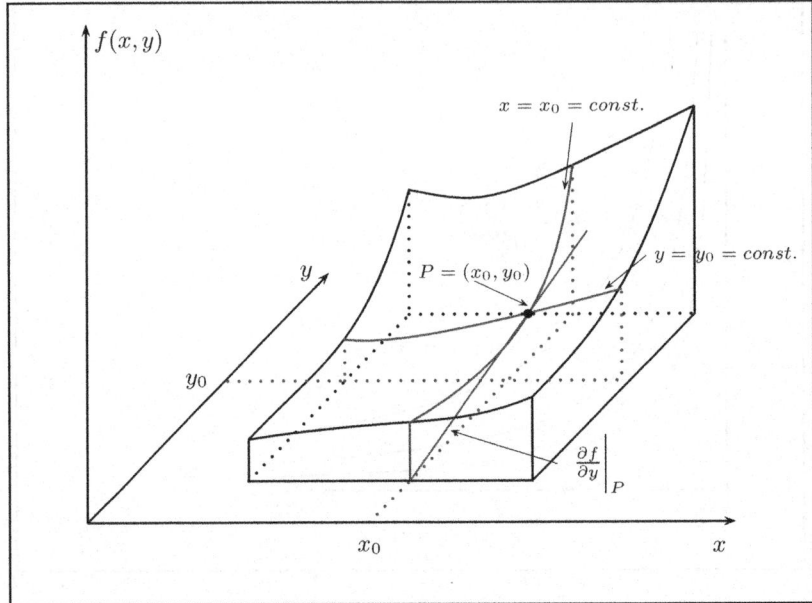

Fig. 6.6 Illustration of the partial derivative

6.2 Differential Calculus for Functions with Several Variables

To investigate the gradient behavior of functions with several variables, we build on the definition of the derivative of a function with one variable. The modifications that are required for this do not depend on the number of variables, but are of a conceptual nature. The graphical illustration for the case of two variables is very helpful (Fig. 6.6).

6.2.1 Partial Derivative

The gradient of a function with several variables in the direction of a coordinate axis x_i results if we only allow changes for this coordinate and keep all other coordinates constant. Accordingly, we define the **partial derivative**:

▶ **Definition (Partial Derivative)** The partial derivative of the function $f(\mathbf{x})=f(x_1, x_2, \ldots, x_n)$ with respect to the variable x_i is the derivative of the function $f(\mathbf{x})$ with respect to the variable x_i, where all variables x_j with $j \neq i$ are considered as constant.

For the partial derivative we use the notation: $\frac{\partial f}{\partial x_i}$.

6.2 Differential Calculus for Functions with Several Variables

Example For the function:
$$f(\mathbf{x}) = x_1^2 + 2x_1 x_2 - 3x_2^2$$

we get the following partial derivatives:

$$\frac{\partial f}{\partial x_1} = 2x_1 + 2x_2, \qquad \frac{\partial f}{\partial x_2} = 2x_1 - 6x_2.$$

For
$$f(\mathbf{x}) = x_1 \cdot e^{ax_2}$$

we get:
$$\frac{\partial f}{\partial x_1} = e^{ax_2}, \qquad \frac{\partial f}{\partial x_2} = ax_1 e^{ax_2}.$$

There are n partial derivatives $\frac{\partial f}{\partial x_i}$; $i = 1, 2, \ldots, n$ of a function $f(\mathbf{x})$. They can be summarized in a vector, the **gradient of the function** f.

▶ **Definition (Gradient)** If there exist the partial derivatives $\frac{\partial f(\mathbf{x})}{\partial x_i}$ of a function $f(\mathbf{x})$ with $\mathbf{x} \in \mathbb{R}^n$ at the point $\mathbf{x}$, then

$$\mathbf{grad} f(\mathbf{x}) = \begin{pmatrix} \frac{\partial f(\mathbf{x})}{\partial x_1} \\ \frac{\partial f(\mathbf{x})}{\partial x_2} \\ \vdots \\ \frac{\partial f(\mathbf{x})}{\partial x_n} \end{pmatrix}$$

is called the gradient of f at the point $\mathbf{x}$.

The gradient of a function of several variables gives the direction of the steepest ascent of the function $f(\mathbf{x})$,[3] the absolute value of the gradient $| \mathbf{grad} f(\mathbf{x}) |$ is a measure of the size of this ascent.

Example We consider the function:
$$f(x_1, x_2) = 1 - x_1^2 - x_2^2.$$

The gradient is the vector
$$\mathbf{grad}\, f(\mathbf{x}) = \begin{pmatrix} -2x_1 \\ -2x_2 \end{pmatrix}.$$

---

[3] This property is shown by first considering the slope in any direction and then checking in which direction the slope is maximized.

For functions that are partially differentiable, we can also consider higher partial derivatives. For practical applications, in particular, the 2nd partial derivatives are of interest. For a partial derivative $\partial f/\partial x_i$ the 2nd derivative can be considered with respect to any variable x_j. If these derivatives exist at the point $\mathbf{x}$, we can consider the so called **Hesse matrix**:[4]

▶ **Definition (Hesse matrix)**

$$\mathbf{H}(\mathbf{x}) = \left(\frac{\partial^2 f(\mathbf{x})}{\partial x_i \partial x_j} \right) = \begin{pmatrix} \frac{\partial^2 f(\mathbf{x})}{\partial x_1 \partial x_1} & \frac{\partial^2 f(\mathbf{x})}{\partial x_1 \partial x_2} & \cdots & \frac{\partial^2 f(\mathbf{x})}{\partial x_1 \partial x_n} \\ \frac{\partial^2 f(\mathbf{x})}{\partial x_2 \partial x_1} & \frac{\partial^2 f(\mathbf{x})}{\partial x_2 \partial x_2} & \cdots & \frac{\partial^2 f(\mathbf{x})}{\partial x_2 \partial x_n} \\ \vdots & \vdots & \vdots & \vdots \\ \frac{\partial^2 f(\mathbf{x})}{\partial x_n \partial x_1} & \frac{\partial^2 f(\mathbf{x})}{\partial x_n \partial x_2} & \cdots & \frac{\partial^2 f(\mathbf{x})}{\partial x_n \partial x_n} \end{pmatrix}.$$

The Hesse Matrix is a square matrix and because of

$$\frac{\partial^2 f(\mathbf{x})}{\partial x_i \partial x_j} = \frac{\partial^2 f(\mathbf{x})}{\partial x_j \partial x_i}$$

also symmetric. With the help of the Hesse matrix, the **convexity** of a function $f(\mathbf{x})$ is examined. This plays a role, as with functions of one variable, in the formulation of sufficient conditions for local extrema (cf. Sect. 6.3).

6.2.2 The Total Differential

We know from the consideration of functions with one variable that the change Δf of the function f in the point x is described by:

$$\Delta f \approx f'(x) \cdot \Delta x.$$

In the limit $\Delta x \to 0$, we obtain the Differential

$$df = f'(x) dx.$$

We now want to consider an expression for functions with several variables.

The total change df of a function $y = f(\mathbf{x})$ consists of the changes that occur in each coordinate direction:

$$df = \sum_{i=1}^{n} df_{x_i}.$$

---

[4] Named after the German mathematician Ludwig Otto Hesse (1811–1874) (see Dieudonné (1985)).

6.2 Differential Calculus for Functions with Several Variables

For the change df_{x_i} in coordinate direction x_i, we can write with the help of the partial derivative:

$$df_{x_i} = \frac{\partial f}{\partial x_i} dx_i,$$

this is called the **partial differential**. Thus, for the total change (the total differential), the following definition results:

▶ **Definition (Total Differential)** The total differential is the sum of all partial differentials.

$$df = \sum_{i=1}^{n} \frac{\partial f}{\partial x_i} dx_i. \qquad (6.1)$$

Example A production function is given by:

$$\mathbf{x}(r_1, r_2) = 15 \cdot r_1^{\frac{1}{2}} \cdot r_2^2$$

with the two input factors r_1 and r_2 and the output $\mathbf{x}$.

The partial change of the production factors results from the change of r_1 by dr_1 or of r_2 by dr_2. The complete differential of the function $x(r_1, r_2)$ now describes the change of the output with simultaneous change of both input factors:

$$\begin{aligned} dx &= \frac{\partial x}{\partial r_1} dr_1 + \frac{\partial x}{\partial r_2} dr_2 \\ &= \frac{15}{2} r_1^{-\frac{1}{2}} r_2^2 dr_1 + 30 r_1^{\frac{1}{2}} r_2 dr_2. \end{aligned}$$

Another application for the total differential results from **implicit functions**. An implicit function has the form $f(\mathbf{x}) = 0$, for example:

$$f(x_1, x_2) = x_1^2 e^{x_2} - x_1 x_2^2 = 0.$$

The total differential provides a way to form derivatives of implicit functions, even if the function cannot be transformed into explicit form. For such implicit functions

$$f(\mathbf{x}) = f(x_1, x_2) = 0 \qquad (6.2)$$

we consider the total differential:

$$df = \frac{\partial f}{\partial x_1} dx_1 + \frac{\partial f}{\partial x_2} dx_2$$

or:

$$\frac{df}{dx_1} = \frac{\partial f}{\partial x_1} + \frac{\partial f}{\partial x_2} \frac{dx_2}{dx_1}.$$

Because of $f(\mathbf{x}) = 0$ follows

$$\frac{df}{dx_1} = 0$$

and thus:

$$\frac{dx_2}{dx_1} = -\frac{\frac{\partial f}{\partial x_1}}{\frac{\partial f}{\partial x_2}}, \quad \text{if } \frac{\partial f}{\partial x_2} \neq 0. \tag{6.3}$$

Example For

$$f(x_1, x_2) = x_1^2 e^{x_2} - x_1 x_2^2 = 0$$

you get the derivative:

$$\frac{dx_2}{dx_1} = -\frac{\frac{\partial f}{\partial x_1}}{\frac{\partial f}{\partial x_2}} = -\frac{2x_1 e^{x_2} - x_2^2}{x_1^2 e^{x_2} - 2x_1 x_2}.$$

In economics, this procedure is used for substitutable production factors. This is the case when a production level remains constant under the variation of two production factors, as for example given by labor and machine usage or in the case of different energy sources.

Let us consider the previous example with the production function:

$$x(r_1, r_2) = 15 \cdot r_1^{\frac{1}{2}} \cdot r_2^2.$$

We ask how resource r_1 can be replaced by resource r_2 to ensure a production x_0. The relation is given by the slope of the isoquants of

$$x(r_1, r_2) = 15 \cdot r_1^{\frac{1}{2}} \cdot r_2^2 = x_0$$

and depends on a certain point $(\tilde{r}_1, \tilde{r}_2)$. For the function

$$f(r_1, r_2) = 15 \cdot r_1^{\frac{1}{2}} \cdot r_2^2 - x_0 = 0$$

we apply Eq. (6.3) to obtain the derivative:

$$\frac{dr_2}{dr_1} = -\frac{\frac{\partial f}{\partial r_1}}{\frac{\partial f}{\partial r_2}} = -\frac{\frac{15}{2} r_1^{-\frac{1}{2}} \cdot r_2^2}{15 r_1^{\frac{1}{2}} \cdot 2 r_2} = -\frac{1}{4} \cdot \frac{r_2}{r_1}.$$

6.3 Extremum Values of Functions with Several Variables

At the point $(\tilde{r}_1, \tilde{r}_2)$ we thus get for the substitution of r_1 by r_2 the relation

$$dr_2 = -\frac{1}{4} \cdot \frac{\tilde{r}_2}{\tilde{r}_1} \cdot dr_1.$$

6.3 Extremum Values of Functions with Several Variables

In the treatment of extremum problems of functions with several variables, we can fall back on the concepts that we introduced when considering functions with one variable, but now we have to use the terms considered in Sect. 6.2. First, we again have to distinguish between extremum problems without (Fig. 6.7 and 6.8) and with boundary conditions (Fig. 6.9).

6.3.1 Extremum without Boundary Conditions

As in Sect. 3.5 we are looking for relative maxima and minima, without taking into account boundary values. For the function $y = f(\mathbf{x})$ we are looking for the vectors $\mathbf{x}_E$, in which $y_E = f(\mathbf{x}_E)$ is locally maximal or minimal. The necessary condition for this is that the gradient in all directions is zero, i.e. that all partial derivatives are zero. With the help of the gradient vector, this can be formulated as follows:

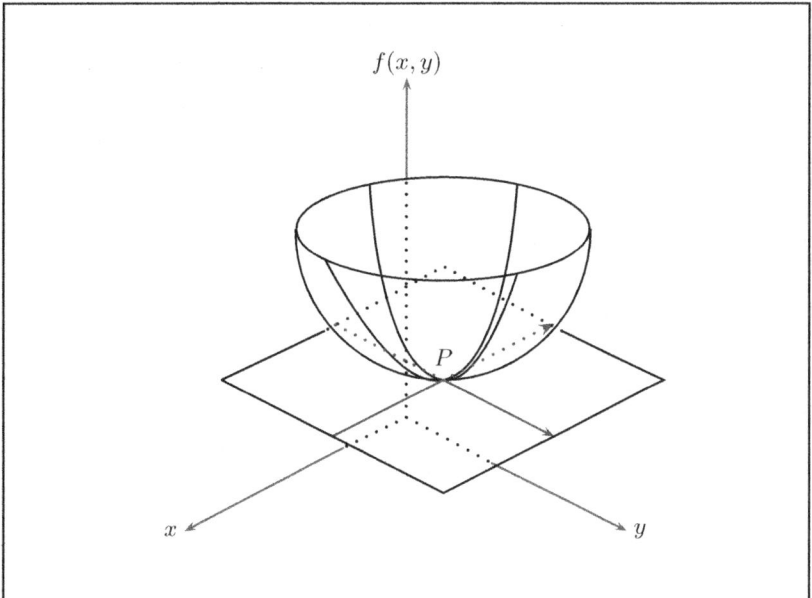

Fig. 6.7 Local minimum of a function with two variables

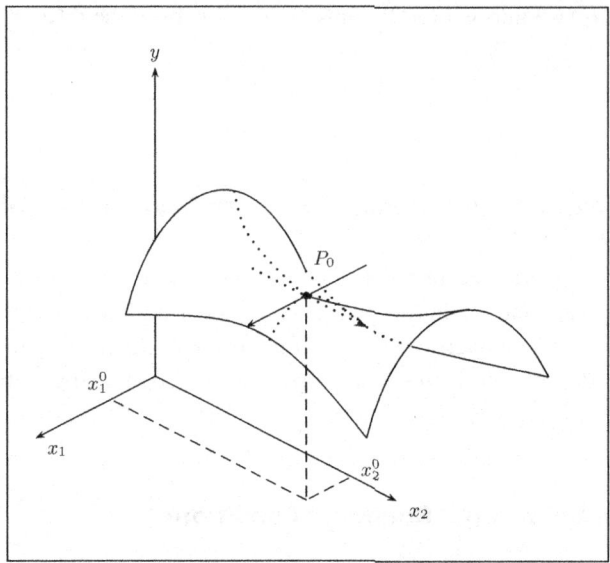

Fig. 6.8 Saddle point

Necessary condition for the existence of local extrema:

If $y=f(\mathbf{x})$ has a local extremum at point $\mathbf{x} = \mathbf{x}_E \in \mathbb{R}^n$, then it must be true:

$$\mathbf{grad} f(\mathbf{x}_E) = \mathbf{0}. \tag{6.4}$$

The condition:

$$\mathbf{grad} f(\mathbf{x}_E) = \mathbf{0}$$

represents a system of n coupled equations. As long as the function $y=f(\mathbf{x})$ only contains quadratic terms, it is a system of linear equations. We have considered efficient solution methods for this in detail in Sect. 5.3.2.

Example We consider the function:

$$f(x_1, x_2) = x_1^2 - 2x_1 x_2.$$

The gradient vector of this function is:

$$\mathbf{grad} f(x_1, x_2) = \begin{pmatrix} \dfrac{\partial f(x_1, x_2)}{\partial x_1} \\ \dfrac{\partial f(x_1, x_2)}{\partial x_2} \end{pmatrix} = \begin{pmatrix} 2x_1 - 2x_2 \\ -2x_1 \end{pmatrix}.$$

6.3 Extremum Values of Functions with Several Variables

The necessary condition for extrema, $\mathbf{grad}\, f(x_1, x_2) \stackrel{!}{=} \mathbf{0}$, leads to the system of linear equations:

$$2x_1 - 2x_2 = 0$$
$$-2x_1 = 0$$

with the solution:

$$x_1 = 0, x_2 = 0.$$

In the general case—if the function $f(\mathbf{x})$ contains higher terms than quadratic—the so-called **gradient method** is an efficient method available. The gradient method can be described as follows (for the search for a local maximum): Starting from any initial point, one looks for the nearest local peak in a mountain landscape. In the initial point, the direction of the steepest ascent is determined and one moves in this direction until it no longer goes uphill. Then one moves from this point again in the direction of the steepest ascent as long as it goes uphill. This is repeated until there is no longer a significant increase in any direction. You are then (approximately) on the local peak. As with the Newton method for determining zeros (see Sect. 3.6.2), the gradient method is an **iterative approximation method**. We need a counter variable k, which counts the iterations of the method and a threshold ε, $\varepsilon > 0$, with which the method is terminated when the gradient is correspondingly small.

Algorithm Gradient Method

Maximize $z = f(\mathbf{x}), \mathbf{x} \in \mathbb{R}^n$

1. *Initialization*:
   Set $k = 0$, choose $\mathbf{x} = \mathbf{x}^{(k)}$ as the initial solution and a termination threshold ε, $(\varepsilon > 0)$. In this first step of the procedure, an initial value must be assumed and the accuracy of the procedure specified.
2. *Termination*:
   If

   $$|\,\mathbf{grad}\, f(\mathbf{x}^{(k)})\,| < \epsilon,$$

   the procedure terminates. The vector $\mathbf{x}^{(k)}$ is the desired maximum. Otherwise continue with step [3].
3. *Determination of the gradient in the steepest direction*:
   Determine $\mu^{(k)} = \mu^{(k)}_{\text{opt}}$ so that the function

   $$g(\mu^{(k)}) = f\left(\mathbf{x}^{(k)} + \mu^{(k)} \mathbf{grad}\, f(\mathbf{x}^{(k)})\right)$$

   is maximized.

In this step, it is specified that just as far as is gone in the direction of the steepest gradient until the function values decrease again. The problem is thereby reduced to the determination of an extremum for a variable ($\mu^{(k)}$) of the function $g(\mu^{(k)})$.

4. *Increasing the iteration step*:
Set $k = k+1$ and

$$\mathbf{x}^{(k)} = \mathbf{x}^{(k-1)} + \mu_{opt}^{(k-1)} \, \mathbf{grad} f(\mathbf{x}^{(k-1)}).$$

In this step, the new approximation for $\mathbf{x}_E$ is determined, which results if one goes from the last approximation in the direction of the steepest gradient by the optimum step width.

Go to step [2].

For the determination of local minima it should be noted that for continuously differentiable functions -grad $f(\mathbf{x})$ indicates the direction of the steepest slope. To determine a local minimum, grad $f(\mathbf{x})$ is therefore replaced by -grad $f(\mathbf{x})$.

Example The function

$$f(x,y) = x^4 + xy + (1+y)^2 \tag{6.5}$$

has a local minimum. The gradient of this function (6.5) is the vector

$$\mathbf{grad} f(x,y) = \begin{pmatrix} 4x^3 + y \\ x + 2(1+y) \end{pmatrix}. \tag{6.6}$$

Starting from an initial solution, the following iterations result from the numerical search for the local minimum:

- Initial solution:

$$\mathbf{x}^{(0)} = \begin{pmatrix} 0 \\ 0 \end{pmatrix}.$$

- 1. Iteration:
With the starting solution we get the gradient (6.6) at the point $\mathbf{x}^{(0)}$ to:

$$\mathbf{grad} f(\mathbf{x}^{(0)}) = \begin{pmatrix} 0 \\ 2 \end{pmatrix}$$

and

$$|\,\mathbf{grad} f(\mathbf{x}^{(0)})\,| = 2.$$

So we calculate the function g:

6.3 Extremum Values of Functions with Several Variables

$$g(\mu^{(0)}) = f\left(x^{(0)} - \mu^{(0)} \operatorname{grad} f(x^{(0)})\right)$$

$$= f\left(\begin{pmatrix}0\\0\end{pmatrix} - \mu^{(0)} \begin{pmatrix}0\\2\end{pmatrix}\right)$$

$$= f\left(\begin{pmatrix}0\\-2\mu^{(0)}\end{pmatrix}\right)$$

$$= (1 - 2\mu^{(0)})^2.$$

This function is minimal at:

$$\frac{dg(\mu^{(0)})}{d\mu^{(0)}} = 2(1 - 2\mu^{(0)}) \stackrel{!}{=} 0$$

or:

$$\mu_{\text{opt}}^{(0)} = \frac{1}{2}.$$

First approximation:

$$x^{(1)} = x^{(0)} - \mu_{\text{opt}}^{(0)} \operatorname{grad} f(x^{(0)})$$

$$= \begin{pmatrix}0\\0\end{pmatrix} - \frac{1}{2}\begin{pmatrix}0\\2\end{pmatrix}$$

$$= \begin{pmatrix}0\\-1\end{pmatrix}.$$

- 2. Iteration: Substitution of the first approximation in Eq. (6.6) yields:

$$\operatorname{grad} f(x^{(1)}) = \begin{pmatrix}-1\\0\end{pmatrix}$$

with:

$$|\operatorname{grad} f(x^{(1)})| = 1.$$

So we calculate the function g in the second iteration:

$$g(\mu^{(1)}) = f\left(x^{(1)} - \mu^{(1)} \operatorname{grad} f(x^{(1)})\right)$$

$$= f\left(\begin{pmatrix}0\\-1\end{pmatrix} - \mu^{(1)} \begin{pmatrix}-1\\0\end{pmatrix}\right)$$

$$= f\left(\begin{pmatrix}\mu^{(1)}\\-1\end{pmatrix}\right)$$

$$= (\mu^{(1)})^4 - \mu^{(1)}.$$

The necessary condition for a minimum is:

$$\frac{dg(\mu^{(1)})}{d\mu^{(0)}} = 4(\mu^{(1)})^3 - 1 \stackrel{!}{=} 0 \quad \Longleftrightarrow \quad \mu^{(1)}_{\text{opt}} = \sqrt[3]{\frac{1}{4}}.$$

Second approximation:

$$\mathbf{x}^{(2)} = \mathbf{x}^{(1)} - \mu^{(1)}_{\text{opt}} \,\mathbf{grad}\, f(\mathbf{x}^{(1)})$$

$$= \begin{pmatrix} 0 \\ -1 \end{pmatrix} - \sqrt[3]{\frac{1}{4}} \begin{pmatrix} -1 \\ 0 \end{pmatrix}$$

$$= \begin{pmatrix} \sqrt[3]{\frac{1}{4}} \\ -1 \end{pmatrix}.$$

Substitution of this approximation in Eq. (6.6) yields:

$$\mathbf{grad}\, f(\mathbf{x}^{(2)}) = \begin{pmatrix} 0 \\ \sqrt[3]{\frac{1}{4}} \end{pmatrix}$$

and

$$|\,\mathbf{grad}\, f(\mathbf{x}^{(1)})\,| = \sqrt[3]{\frac{1}{4}} \approx 0{,}63.$$

The procedure converges, because $|\,\mathbf{grad}\, f(\mathbf{x})\,|$ decreases. For acceptable approximations ($\varepsilon < 0{,}05$) here still some steps would be necessary.

Analogous to functions of one variable, the condition

$$\mathbf{grad}\, f(\mathbf{x}) = \mathbf{0}$$

is only a necessary condition. For functions of several variables this condition is also fulfilled for saddle points which do not represent local extrema. As indicated in Fig. 6.8 saddle points have the property that they have horizontal tangents in all directions. However, if you move away from the point in different directions, you will get larger or smaller function values depending on the direction. Therefore, neither a maximum nor a minimum can occur here. In summary, local extrema and saddle points are also referred to as **stationary points**.

As a **sufficient condition** for the existence of a local extremum, we must consider the curvature behavior or the convexity of the function $z = f(\mathbf{x})$. This allows us to distinguish between saddle points and extrema. For this purpose, we consider the second (partial) derivatives of $z = f(\mathbf{x})$ and state the sufficient conditions without proof (see Tietze (2015)). First, we consider the special case of functions with two variables

$$y = f(x_1, x_2).$$

6.3 Extremum Values of Functions with Several Variables

A differentiable function $y = f(x_1, x_2)$ has a stationary point at $P = (\tilde{x}_1, \tilde{x}_2) \in \mathbb{R}^2$ if the following conditions are met:

$$\left.\frac{\partial f}{\partial x_1}\right|_P = \left.\frac{\partial f}{\partial x_2}\right|_P = 0.$$

This is a **local maximum**, if in P the following conditions are fulfilled:

$$\frac{\partial^2 f}{\partial x_1^2} < 0, \frac{\partial^2 f}{\partial x_2^2} < 0 \quad \text{and} \quad \frac{\partial^2 f}{\partial x_1^2} \cdot \frac{\partial^2 f}{\partial x_2^2} - \left(\frac{\partial^2 f}{\partial x_1 \partial x_2}\right)^2 > 0, \qquad (6.7)$$

and a **local minimum**, if in P it holds:

$$\frac{\partial^2 f}{\partial x_1^2} > 0, \frac{\partial^2 f}{\partial x_2^2} > 0 \quad \text{and} \quad \frac{\partial^2 f}{\partial x_1^2} \cdot \frac{\partial^2 f}{\partial x_2^2} - \left(\frac{\partial^2 f}{\partial x_1 \partial x_2}\right)^2 > 0. \qquad (6.8)$$

If the expression

$$\frac{\partial^2 f}{\partial x_1^2} \cdot \frac{\partial^2 f}{\partial x_2^2} - \left(\frac{\partial^2 f}{\partial x_1 \partial x_2}\right)^2 < 0,$$

then there is a saddle point. If

$$\left.\frac{\partial^2 f}{\partial x_1^2} \frac{\partial^2 f}{\partial x_2^2} - \left(\frac{\partial^2 f}{\partial x_1 \partial x_2}\right)^2 \right|_P = 0,$$

then the stationary point P cannot be characterized any further by means of the 2nd derivatives.

Example We consider the function

$$f(x_1, x_2) = x_1^2 - \frac{1}{2}x_1 x_2 + x_2^2.$$

The necessary condition $\partial f / \partial x_i = 0$; $i = 1, 2$ leads to the system of equations

$$2x_1 - \frac{1}{2}x_2 = 0$$

$$2x_2 - \frac{1}{2}x_1 = 0,$$

which leads to the stationary point $P = (0,0)$. To decide whether in P there is a local minimum, a maximum or a saddle point, we form the second derivatives:

$$\frac{\partial^2 f}{\partial x_1^2} = 2, \quad \frac{\partial^2 f}{\partial x_2^2} = 2, \quad \frac{\partial^2 f}{\partial x_1 \partial x_2} = -\frac{1}{2}.$$

Since

$$\frac{\partial^2 f}{\partial x_1^2} > 0, \quad \frac{\partial^2 f}{\partial x_2^2} > 0$$

and

$$\frac{\partial^2 f}{\partial x_1^2} \cdot \frac{\partial^2 f}{\partial x_2^2} - \left(\frac{\partial^2 f}{\partial x_1 \partial x_2}\right)^2 = \frac{15}{4} > 0$$

the function f has in $P = (0,0)$ a local minimum.

For functions with more than two variables, the necessary condition can be formulated using the **Hesse matrix** in the following form:[5]

A stationary point P is:

Relative maximum, if in P all eigenvalues of the Hesse matrix are negative.
Relative minimum, if in P all eigenvalues of the Hesse matrix are positive.
If the eigenvalues of the Hesse matrix have different signs, then P is a **saddle point**. The above conditions for functions with two variables are a special case of these generally valid conditions.

In summary, we present necessary and sufficient conditions for local extrema with respect to functions with one and several variables.[6]

	$y = f(x)$	$y = f(\mathbf{x})$
necessary Condition	$f'(x) = 0$	$\mathbf{grad} f(\mathbf{x}) = \mathbf{0}$
sufficient Condition	$f''(x) > 0$ Minimum	All eigenvalues $\mathbf{H}(\mathbf{x}) > 0$
		$f(\mathbf{x})$ convex from above
	$f''(x) < 0$ Maximum	All eigenvalues $\mathbf{H}(\mathbf{x}) < 0$
		$f(\mathbf{x})$ concave from above

6.3.2 Extremum Values with Boundary Conditions

The consideration of the boundary does not pose any fundamental difficulties for functions with one variable. You just have to calculate the two function values at the boundary of the definition area and compare them with the function values of the local extrema.

---

[5] See, for example, Arens et al. (2018), Sect. 24.6, Goebbels and Ritter (2011), Sect. 4.3 or Stoeppler (1982).
[6] As in the example $y = x^6$ given in Sect. 3.5 there are also special functions in the case of several variables for which the consideration of the 2nd derivative is not sufficient. Then one has to examine the environment of f in the stationary point.

6.3 Extremum Values of Functions with Several Variables

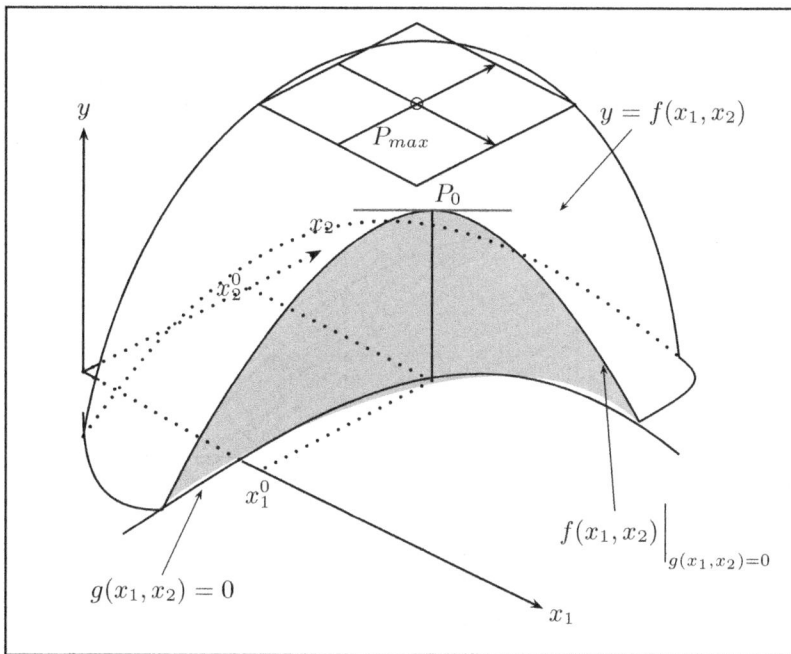

Fig. 6.9 Geometrical illustration of a function with two variables and the constraint $g(x_1, x_2) = 0$. Without the constraint, the function $f(x_1, x_2)$ has a maximum at point P_{max}. The consideration of the constraint $g = 0$ restricts the area $y = f(x_1, x_2)$ to the curve $f(x_1, x_2)|_{g=0}$. On this curve, the maximum is at point P_0

The general treatment of extremum problems with boundary conditions exceeds the scope of this introduction.[7] We restrict ourselves here to the case that the boundary conditions are given as equations (in the general case one has to deal with inequalities) and formulate the necessary conditions for this. We first restrict ourselves to two variables and consider the function:

$$y = f(x_1, x_2)$$

with the **boundary condition**
$$g(x_1, x_2) = 0.$$

In Fig. 6.9 the geometric interpretation of taking into account the boundary condition is shown. For the special case that the boundary condition can be written as an explicit function $x_2 = x_2(x_1)$, a variable substitution also helps (elimination method). In the following we want to consider the general approach of the **Lagrange multipliers**.

---

[7] See Arens et al. (2018), Sect. 35.2 or Goebbels and Ritter (2011), Sect. 4.3.2.

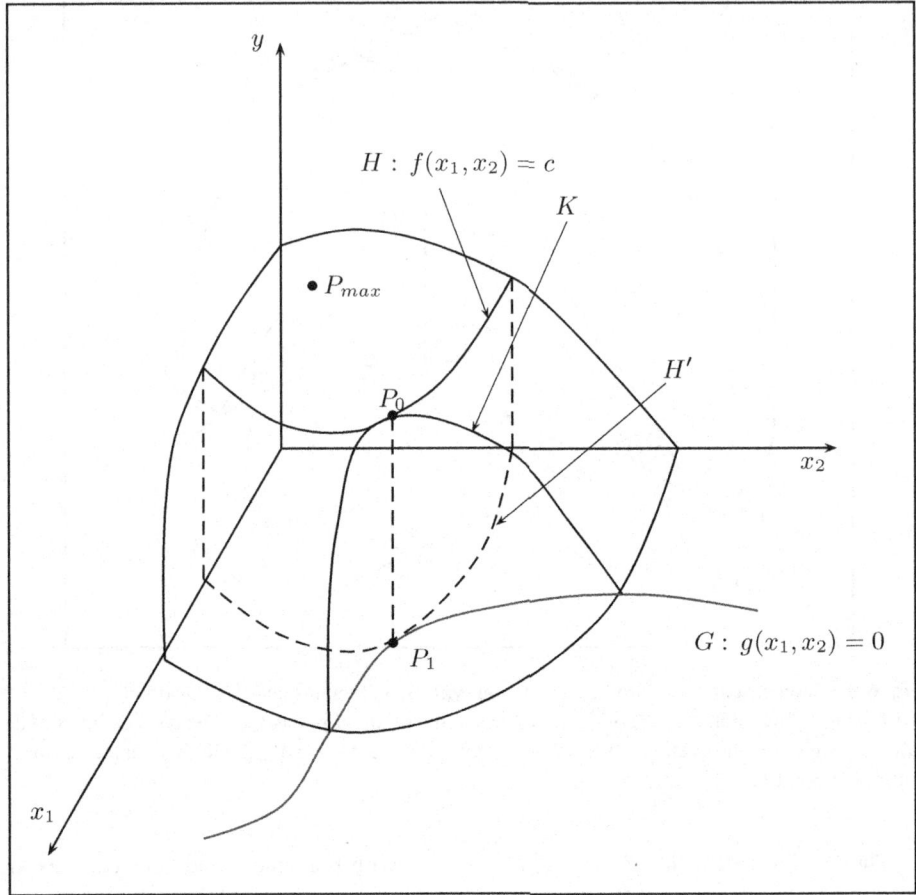

Fig. 6.10 To derive the method of Lagrange multipliers

Method of Lagrange Multipliers
The idea underlying the method of Lagrange multipliers is illustrated in Fig. 6.10. Without the constraint, the function $f(x_1,x_2)$ the maximum in a point P_{max}. Through the boundary condition

$$g(x_1, x_2) = 0$$

the considered area

$$y = f(x_1, x_2)$$

is reduced to a space curve K whose projection into the $x_1 - x_2$-plane is the curve G (G is just the side condition $g(x_1, x_2)=0$). Now the maximum of the space curve K is sought. In Fig. 6.10 this is the point P_0. P_0 is the point of contact of the space curve K and a height line H which can be described by

6.3 Extremum Values of Functions with Several Variables

$$H : f(x_1, x_2) = c$$

The sought coordinates of the maximum P_0 (under the side condition $g=0$) result from the point P_1 in the $x_1 - x_2$-plane. P_1 is—as the figure shows—in turn the point of contact of the curves G and H'. The curve H' describes the projection of the height line H into the $x_1 - x_2$-plane. The touching of the two curves at the point P_1 means that the gradients of both curves match in this point. We obtain the gradients of these curves from the derivatives and these in turn from the implicit form of the functions.

$$g(x_1, x_2) = 0 \quad \text{rsp.} \quad f(x_1, x_2) - c = 0$$

according to Eq. (6.3):

$$\frac{dx_2}{dx_1} = -\frac{\frac{\partial g}{\partial x_1}}{\frac{\partial g}{\partial x_2}} = -\frac{\frac{\partial f}{\partial x_1}}{\frac{\partial f}{\partial x_2}}. \tag{6.9}$$

Eq. (6.9) means that the partial derivatives of f and g are proportional to each other. With the proportionality factor $(-\lambda)$ it must therefore hold:

$$\frac{\partial f}{\partial x_1} = -\lambda \frac{\partial g}{\partial x_1} \quad \text{and} \quad \frac{\partial f}{\partial x_2} = -\lambda \frac{\partial g}{\partial x_2}$$

or

$$\frac{\partial f}{\partial x_1} + \lambda \frac{\partial g}{\partial x_1} = 0 \tag{6.10}$$

and

$$\frac{\partial f}{\partial x_2} + \lambda \frac{\partial g}{\partial x_2} = 0. \tag{6.11}$$

The factor λ is called the **Lagrange multiplier**.[8]

You can also get these equations if you form the so-called **Lagrange function** from f and g in the form:

$$\mathcal{L}(x_1, x_2, \lambda) = f(x_1, x_2) + \lambda \cdot g(x_1, x_2). \tag{6.12}$$

The necessary conditions can be formed from this Lagrange function by the partial derivatives:

$$\frac{\partial \mathcal{L}}{\partial x_1} = 0 \quad \text{and} \quad \frac{\partial \mathcal{L}}{\partial x_2} = 0. \tag{6.13}$$

---

[8] Named after the Italian-French mathematician Joseph Louis Lagrange (1736–1813).

The derivative of the Lagrange function with respect to the parameter λ reproduces the boundary condition:

$$\frac{\partial \mathcal{L}}{\partial \lambda} = 0 \quad \Longleftrightarrow \quad g(x_1, x_2) = 0. \tag{6.14}$$

From Eqs. (6.12) and (6.14), the coordinates of the point P_0 can be determined at which the function $f(x_1, x_2)$ has an extremum under the boundary condition $g(x_1, x_2) = 0$.

Examples
1. We consider the function

$$f(x_1, x_2) = x_1^2 - 2x_1 x_2 \tag{6.15}$$

with the boundary condition

$$x_2 = 2x_1 - 6. \tag{6.16}$$

Now we are looking for the point (or points) from $\mathbb{R}^2$, at which the function (6.15) has an extremum, taking into account the boundary condition (6.16). For this we consider the Lagrange function (cf. Eq. (6.12)):

$$\mathcal{L} = f(x_1, x_2) + \lambda g(x_1, x_2)$$
$$= x_1^2 - 2x_1 x_2 + \lambda(x_2 - 2x_1 + 6).$$

We form the partial derivatives:

$$\frac{\partial \mathcal{L}}{\partial x_1} = 2x_1 - 2x_2 - 2\lambda \stackrel{!}{=} 0 \tag{6.17}$$

$$\frac{\partial \mathcal{L}}{\partial x_2} = -2x_1 + \lambda \stackrel{!}{=} 0 \tag{6.18}$$

$$\frac{\partial \mathcal{L}}{\partial \lambda} = x_2 - 2x_1 + 6 \stackrel{!}{=} 0 \tag{6.19}$$

Eqs. (6.17)–(6.19) form a linear system. From Eqs. (6.18) and (6.19) it follows:

$$\lambda = 2x_1 \quad \text{and} \quad x_2 = 2x_1 - 6.$$

Inserting in (6.17) gives:

$$x_1 = 2$$

and thus: $x_2 = -2$. This makes the point $P = (2, -2) \in \mathbb{R}^2$ an extremum of the function (6.15) under the boundary condition (6.16).

2. Determination of the **minimum cost combination**
   There is a production function $x(\mathbf{r})$. The costs arise from the prices of the input factors $\mathbf{r}$, the combination of input factors is sought with which a given output can be generated at minimum cost. For a Cobb-Douglas production function

6.3 Extremum Values of Functions with Several Variables

$$x(r_1, r_2) = c \cdot r_1^\alpha \cdot r_2^\beta$$

with the cost function

$$K(r_1, r_2) = p_1 \cdot r_1 + p_2 \cdot r_2,$$

the output x_0: Minimize $K(r_1, r_2)$ under the condition $c \cdot r_1^\alpha \cdot r_2^\beta = x_0$. The Lagrange function is

$$\mathcal{L}(r_1, r_2, \lambda) = p_1 \cdot r_1 + p_2 \cdot r_2 + \lambda \cdot \left(x_0 - c \cdot r_1^\alpha \cdot r_2^\beta\right).$$

Thus:

$$\frac{\partial \mathcal{L}}{\partial r_1} = p_1 - \lambda \cdot c \cdot \alpha \cdot r_1^{\alpha-1} \cdot r_2^\beta \stackrel{!}{=} 0, \tag{6.20}$$

$$\frac{\partial \mathcal{L}}{\partial r_2} = p_2 - \lambda \cdot c \cdot \beta \cdot r_1^\alpha \cdot r_2^{\beta-1} \stackrel{!}{=} 0, \tag{6.21}$$

$$\frac{\partial \mathcal{L}}{\partial \lambda} = x_0 - c \cdot r_1^\alpha \cdot r_2^\beta \stackrel{!}{=} 0. \tag{6.22}$$

Eqs. (6.20) and (6.21) are solved for λ and set equal. This results in

$$r_2 = \frac{p_1 \cdot \beta}{p_2 \cdot \alpha} \cdot r_1.$$

In the r_1-r_2 diagram, the cost-minimizing input factors for a Cobb-Douglas production function lie on a straight line (see Fig. 6.11). For r_1, using Eq. (6.22), we get:

$$r_1 = \left(\frac{x_0}{c}\right)^{1/(\alpha+\beta)} \cdot \left(\frac{p_2 \cdot \alpha}{p_1 \cdot \beta}\right)^{\beta/(\alpha+\beta)}.$$

The method of Lagrange multipliers can be generalized to n variables and m constraints: For a function $y = f(\mathbf{x})$ with $\mathbf{x} \in \mathbb{R}^n$ with the constraints

$$g_j(\mathbf{x}) = 0, \quad j = 1, 2, \ldots, m,$$

the relative extrema satisfy the necessary conditions

$$\frac{\partial \mathcal{L}}{\partial x_i} = 0 \quad \text{for } i = 1, 2, \ldots, n$$

$$\frac{\partial \mathcal{L}}{\partial \lambda_j} = 0 \quad \text{for } j = 1, 2, \ldots, m$$

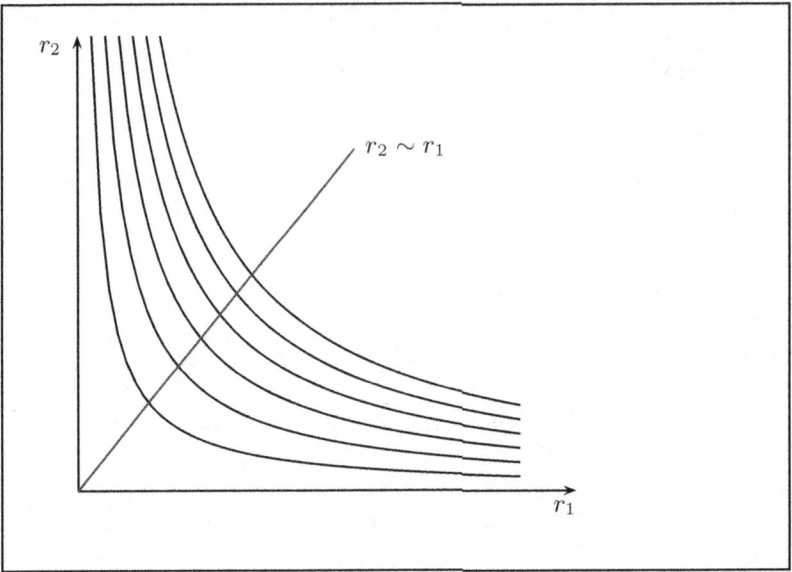

Fig. 6.11 Isoquants of the Cobb-Douglas production function and the determination of minimum cost input factors

with the Lagrange function

$$\mathcal{L} = \mathcal{L}(\mathbf{x}, \lambda) = f(\mathbf{x}) + \sum_{j=1}^{m} \lambda_j g_j(\mathbf{x}).$$

Interesting applications arise when the boundary conditions are in the form of **inequalities**. We refer here to the relevant literature on operations research (see Domschke et al. (2015) or Arens et al. (2018), Chap. 35).

6.4 Exercises

Short solutions to the following exercises can be found in the appendix.

6.1 Calculate the partial derivatives of the following functions:

(a) $f(\mathbf{x}) = x_1^3 + x_1 e^{x_2} + x_2^2 \cdot x_3$
(b) $f(\mathbf{x}) = x_1 \cdot \ln(x_2 x_3) - \ln(x_1 + x_2)$

6.4 Exercises

6.2 Form the gradient of the function

$$f(x,y) = 16 - (x-1)^2 - \left(y - \frac{1}{2}\right)^2.$$

Draw a contour map for $f(x, y)$ ($x-y$-diagram with $f(x, y) = const.$). Enter the gradient for

$$\begin{pmatrix} x_1 \\ y_1 \end{pmatrix} = \begin{pmatrix} 0 \\ 0 \end{pmatrix} \quad \text{and} \quad \begin{pmatrix} x_2 \\ y_2 \end{pmatrix} = \begin{pmatrix} 1 \\ 0 \end{pmatrix}.$$

6.3 Given is the output function

$$x(r_1, r_2, r_3) = 3r_1^2 \cdot \sqrt{r_2} + 5r_1 r_2 r_3^{\frac{1}{3}}.$$

Calculate the total differential of $x(r_1, r_2, r_3)$. How does a change of the input function at the point (1, 1, 1) affect if r_1 and r_2 are increased by 0.1 units and r_3 is decreased by 0.2 units?

6.4 Given the production function

$$x(r_1, r_2) = 2\sqrt{r_1} \cdot \sqrt[3]{r_2}.$$

(a) How large must the factor input of r_1 be in order to ensure a production of $x_0 = 216$ units if $r_2 = 27$?
(b) How can a reduction of the first factor by one unit be compensated by the second factor?

6.5 Show that the sufficient condition for extrema of functions for two variables follows from the general condition for the eigenvalues of the Hessian matrix.

6.6 Given the function

$$f(x, y) = 4x - 2x^2 + 2y - 6y^2 + 2xy$$

(a) Check whether $f(x, y)$ is a concave function from above.
(b) Determine an approximate maximum of the function using the gradient method.
Note: Start at the point $\mathbf{x} = \begin{pmatrix} 0 \\ 0 \end{pmatrix}$ and break the process if $|\mathbf{grad}\,(f(\mathbf{x})| < 0{,}5$ is.
(c) Compare the approximately determined solution with the exact solution, which can be easily found here when the LGS grad $f(\mathbf{x}) = \mathbf{0}$ is solved.
(d) The boundary condition $y = x + 3$ is now to be taken into account in addition. Calculate the maximum of $f(x, y)$, which results from this boundary condition, first by substituting a variable. Then consider how the maximum is obtained using the Lagrange formalism.

6.7 Given the implicit function

$$f(x,y) = x^{-2}e^{-y} + x^2 y^3 = 0.$$

What is the derivative $y'(x)$?

6.8 Given the function

$$f(x) = x^2 - 3x + 3.$$

The point P of the curve that is closest to the origin is sought.

6.9 The function

$$f(x,y) = x^2 + 2xy$$

is subject to the boundary condition

$$g(x,y) = -\frac{3}{2}x + 3y + 6 = 0.$$

Determine the extreme values of the function $f(x, y)$ under the above boundary condition

(a) by the elimination method
(b) by applying the method of Lagrange multipliers.

6.10 The output x is given in dependence of the input factors r_1 and r_2 by a Cobb-Douglas production function:

$$x(r_1, r_2) = c \cdot r_1^\alpha \cdot r_2^\beta.$$

How are the resources to be set so that the output is maximized if the costs of resource use are fixed in the form:

$$K(r_1, r_2) = h_1 r_1 + h_2 r_2 = b?$$

6.11 In a metal processing plant, cylindrical containers made of sheet metal are produced. These containers are closed on both sides because they are used for transporting liquids.

(a) Calculate the height h and the radius r so that the volume is maximized. 2 m² of sheet metal is available for each container.
(b) Show that the solution is unique and calculate the maximum volume.
(c) What must h and r be for a volume of 1 m³ content and minimum sheet metal consumption?
(d) Which geometric body has a smaller, optimal sheet metal consumption with the same volume of 1 m³?

- What are the dimensions of this body?
- How many square meters of sheet metal are necessary?
- What percentage saving is this compared to the cylinder considered above?

References

Arens T., Hettlich F., Karpfinger Ch., Kockelkorn U., Lichtenegger K., Stachel H. (2018): Mathematik, 4. Auflage, Spektrum Akademischer Verlag, Heidelberg.

Dieudonné J. (1985): Geschichte der Mathematik 1700–1900, Vieweg, Braunschweig.

Domschke W., Drexl A., Klein R., Scholl A. (2015): Einführung in Operations Research, 9. Auflage, Springer, Berlin.

Dyke P. (2018): Two and Three Dimensional Calculus with Applications in Science and Engineering. Wiley, Hoboken, NJ.

Financial Mathematics   7

Learning Objectives (This Chapter Provides)

- the most important concepts of financial mathematics
- a consideration of interest and compound interest rate
- how to distinguish between nominal and effective interest rate
- some considerations in connection with annuity loans ◄

7.1 Interest Calculation

In the context of interest calculation, we consider the following variables:

1. K_n is called **final value** and refers to the capital accumulated at the end of the interest period.
2. K_0 is called **present value**, which is the invested capital at the beginning of the investment.
3. p is called **interest rate** in percent per interest period.
4. n is the number of interest periods.
5. The size Z_n indicates the **interest amount** after n interest periods.

7.1.1 Simple Interest

With simple interest—also called **linear interest**—there is no interest surcharge during the capital lending period.

▶ **Definition (Linear Interest)** If a capital K_0 € is at $p\%$ interest (per interest period) for a total of n periods, then for the total amount of interest to be paid at the end of period n:

$$Z_n = \frac{K_0 \cdot p \cdot n}{100}. \tag{7.1}$$

The present value results in:

$$\begin{aligned} K_n &= K_0 + Z_n \\ &= K_0 \cdot \left(1 + \frac{p \cdot n}{100}\right) \\ &= K_0 \cdot (1 + i \cdot n) \end{aligned}$$

with the **percentage**:

$$i = \frac{p}{100}.$$

In financial mathematics, the following **periods** are used: In Germany the year is divided into 12 equal months with 30 days each. This makes the interest year 360 days. In addition, the following conventions apply:

- If the interest surcharge is made at the end of February, 28 or 29 days are calculated.
- When determining durations, the first day is not counted, the last day is counted.
- If the interest rate p is missing a time specification, the specification always refers to one year.

Example An application of linear interest rate calculation from business practice is the so-called **supplier credit**. This is about the following question: For the payment of a goods delivery, the payment conditions apply:

3% discount for payment within 10 days, otherwise full payment of the invoice amount within 30 days.

For simplicity, we assume that it is an invoice amount of 100,– €. The payment conditions are shown in Fig. 7.1.

If the customer of the supplier does **not** take up the offer *payment with discount to the earlier date*, the supplier of the goods grants him a credit in the amount of the invoice amount reduced by the discount—here 97,– €—which the customer has to pay back 20 days later in the amount of 100, – €.

Assuming the customer pays the earlier due datea and finances the earlier payment of 97, – € at 18% pa for the duration of the discount period 20 days. Then his debts after 20 days are:

$$97 \cdot \left(1 + 0.18 \cdot \frac{20}{360}\right) € = 97{,}97 \ €.$$

7.1 Interest Calculation

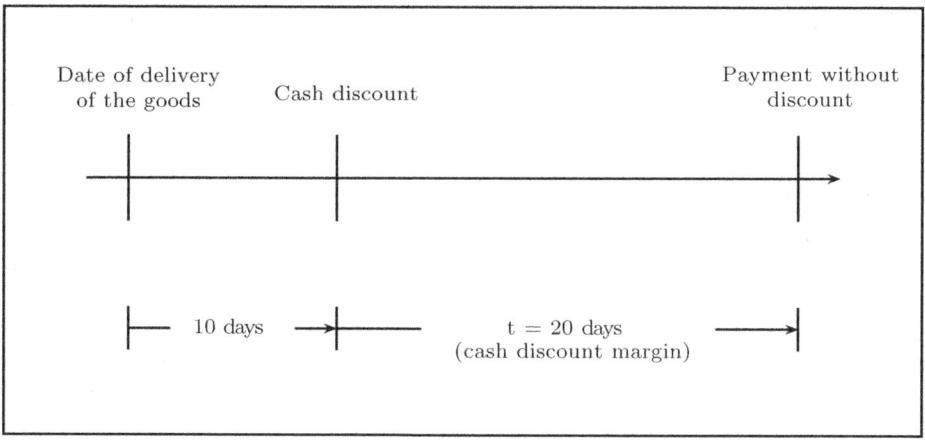

Fig. 7.1 The principle of the discount

This amount would now have to be returned to the buyer's credit bank. If he had used the supplier credit and paid at a later date, 100,– € would have been due. This becomes even clearer when one considers which annual interest rate the supplier credit actually corresponds to.

So what is sought is the interest rate p at which 97,– € in 20 days grows to 100,– €. The corresponding equation provides:

$$97 \cdot \left(1 + i \cdot \frac{20}{360}\right) = 100.$$

If you solve this equation for p, then follows:

$$i = \left(\frac{100}{97} - 1\right) \cdot \frac{360}{20} = 0{,}5567.$$

The supplier credit corresponds to an interest rate of $p = 55.67\%$ when annualized.

7.1.2 Compound Interest

The characteristic of linear interest is that no interest is charged within the period under consideration. If linear interest is agreed, the capital and the interest incurred are only credited at the end of the period under consideration.

Another principle is the **compound interest**. Within the term of the loan, there are one or more interest payment dates, on which the interest accrued up to this point is added to the capital. This amount forms the capital that still earns interest. This procedure of compounding interest is used in many areas such as **investment accounting**, **financing** and **insurance**.

If the interest rate is $p\%$, then the simple annual interest is $\frac{k \cdot p}{100}$. The principal of K_0 will then grow over the course of a year to K_1:

$$K_1 = K_0 \left(1 + \frac{p}{100}\right)$$

with the **interest factor** q:

$$q = 1 + \frac{p}{100} = 1 + i.$$

This gives:

$$\begin{aligned}\text{after 1 year}: \quad & K_1 = K_0 \cdot q \\ \text{after 2 years}: \quad & K_2 = K_1 \cdot q = K_0 \cdot q^2 \\ \text{after 3 years}: \quad & K_3 = K_2 \cdot q = K_0 \cdot q^3 \\ & \vdots \\ \text{after n years}: \quad & K_n = K_{n-1} \cdot q = K_0 \cdot q^n.\end{aligned}$$

It follows that the principal K_0 will grow to the amount K_n in n years, if the interest accrued annually is added to the capital.

The **compound interest formula** is:

▶ **Definition (Compound Interest Formula)**

$$K_n = K_0 \cdot q^n. \tag{7.2}$$

The factor q^n is called **compounding factor**.

Compound interest is an example of **exponential growth**. With exponential growth, the relationship holds:

$$\frac{K_n}{K_{n-1}} = q = const.$$

Exponential growth also occurs, for example, in the growth of living things, as long as no limiting factors act restrictively.

The compound interest formula links the four variables K_n, K_0, q and n by

$$K_n = K_0 \cdot q^n.$$

Solving for present value yields:

$$K_0 = \frac{K_n}{q^n}. \tag{7.3}$$

7.1 Interest Calculation

For the interest factor q we get:

$$q = \sqrt[n]{\frac{K_n}{K_0}}. \tag{7.4}$$

The number of interest periods is calculated according to:

$$n = \frac{\ln\left(\frac{K_n}{K_0}\right)}{\ln q}. \tag{7.5}$$

7.1.3 Annuity Calculation

In the following, we consider the case that over an entire duration of n periods, a regular payment of a rate r is made. We calculate the final value of the investment at an interest rate of p percent, if the interest is added at the end of each period. This is the case, for example, with capital life insurance or the saving of building savings contracts. With $q = 1 + p/100$ follows:

After the first period and directly after the second payment, the credit is $r + r \cdot q$, because the credit consists of the first payment, the interest from the first payment and the second payment.

After the second period and directly after the third payment, the credit is $r + r \cdot q + r \cdot q^2$.

After the third period and directly after the fourth payment, the credit is $r + r \cdot q + r \cdot q^2 + r \cdot q^3$.

$\vdots$

After the n-th period and immediately after the $n+1$-th payment, the credit is: $r + r \cdot q + \cdots + r \cdot q^n$.

Thus, the total value at the n-th payment is given by the following sum, the so-called annuity formula:

$$G_n = r + qr + q^2 r + \cdots + q^{n-1} r$$

$$= r \cdot \sum_{j=1}^{n} q^{j-1}$$

$$= r \cdot \frac{q^n - 1}{q - 1}.$$

▶ **Definition (Annuity Formula)** The annuity formula

$$G_n = r \cdot \frac{q^n - 1}{q - 1} \tag{7.6}$$

gives the capital that, after n-fold payment of a rate r, with an interest rate of $q = 1 + \frac{p}{100}$, is accumulated.

For the derivation of the annuity formula, we have used the sum formula of the **geometric series** see derivation of Eq. (2.12).

Example How many years does it take for a saver to reach a balance of 100,000 € with the last payment if 9600 € are paid in annually and a credit interest of 1 % is paid?
According to the annuity formula Eq. (7.6)

$$G_n = r \cdot \frac{q^n - 1}{q - 1}$$

we get

$$\frac{G_n}{r} \cdot (q - 1) = q^n - 1$$

or

$$\frac{G_n}{r} \cdot (q - 1) + 1 = q^n.$$

Logarithmizing this equation gives:

$$\frac{\ln\left(\frac{G_n}{r} \cdot (q - 1) + 1\right)}{\ln q} = n.$$

Inserting the values $G_n = 100.000$ €; $r = 9.600$ €, $q = 1 + p/100 = 1,01$ leads to

$$n \approx 9,95 \text{ years.}$$

7.1.4 Yearly Interest

It is often the case—for example with securities—that interest is not calculated at the end of a year, but semi-annually, quarterly or at other intervals, and added to the capital. In this case, one speaks of a **yearly interest calculation** or **yearly interest**. When repaying a loan, the interest is usually not calculated at the end of the year, but semi-annually, accordingly the interest debt is added to the loan.

With a fixed interest rate and a subdivision of the year into m equal intervals, the compound interest formula Eq. (7.2)

$$K_n = K_0(1 + i)^n$$

is to be modified as follows:

$$K_n = K_0\left(1 + \frac{i}{m}\right)^{n \cdot m}.$$

7.1 Interest Calculation

This change takes into account the fact that the number of surcharges will be increased by a factor of m per year, with an interest rate that is reduced by a factor of m. With semi-annual interest, p/m is called the **relative interest rate** or the **relative intra-year interest**. i is called the **nominal annual interest rate**.

Example An initial capital of $k = 6000$ € is invested for 5 years at 6%. Calculate the final capital for monthly and annual interest.

For monthly interest calculation:

Annual interest rate: 6%
Number of periods: $m = 12$
Relative intra-year interest p/m: 6% / 12 = 0.5%.
Total number of periods: 5 * 12 = 60
End capital:

$$K_5 = K_0 \left(1 + \frac{p/m}{100}\right)^{mn} = 6.000 \text{ €} \cdot (1{,}005)^{60} = 8.093 \text{ €}.$$

In the case of an annual interest calculation, the following end capital results:

$$K_5 = K_0 \cdot q^5 = 6.000 \text{ €} \cdot (1{,}06)^5 = 8.029 \text{ €}$$

In general, it can be said that with a continued increase in the number of annual periods, the interest increases.

In order to be able to compare different interest conditions for an investment or a loan consistently, the term **effective interest** is introduced in financial mathematics.

In the case of intra-year interest, the **effective annual interest rate** i_{eff} describes the interest rate that would apply in the first year if there were no intra-year interest.

▶ **Definition (Effective Interest Rate with Intra-Year Interest)** The relationship between the nominal interest rate i and the effective annual interest rate is given by

$$\left(1 + \frac{i}{m}\right)^m = 1 + i_{\text{eff}}, \tag{7.7}$$

if m is the number of periods into which the interest year is divided.

Example A nominal interest rate of 2% per year is agreed for a loan with monthly interest rate calculation. How high is the effective annual interest rate in the first year?

From

$$\left(1 + \frac{i}{m}\right)^m = 1 + i_{\text{eff}}$$

follows:

$$i_{\text{eff}} = \left(1 + \frac{i}{m}\right)^m - 1$$

$$= \left(1 + \frac{0{,}02}{12}\right)^{12} - 1$$

$$\approx 0{,}0201.$$

The effective annual interest rate is therefore 2.01%.

The limiting case $m \to \infty$ means permanent interest. This is also called **continuous interest**. The consideration is a good example of the role of the e-function (cf. Sect. 2.2.8) in growth processes. We start from

$$K_n = K_0 \left(1 + \frac{i}{m}\right)^{m \cdot n} = K_0 \left(1 + \frac{1}{\frac{m}{i}}\right)^{\frac{m}{i} \cdot i \cdot n}$$

with $\frac{m}{i} = x$

If we now carry out the limiting case $m \to \infty$ or $x \to \infty$, we obtain:

$$\lim_{x \to \infty} K_0 \left(1 + \frac{1}{x}\right)^{x \cdot i \cdot n} = K_0 \left(\underbrace{\lim_{x \to \infty} \left(1 + \frac{1}{x}\right)^x}_{e}\right)^{i \cdot n} = K_0 \cdot e^{i \cdot n}.$$

In the case of continuous interest, K_n is:

$$K_n = K_0 \cdot e^{i \cdot n}.$$

7.2 Repayment Calculation

The **repayment calculation** is the mathematical basis for every debt repayment of a loan, credit or mortgage. It is therefore used in the context of the **investment**, but also of interest to the private debtor.

In all debt repayment transactions, there are:

- Services = payments of the lender (= creditor)
- and consideration (= payments) of the borrower (= debtor)

A characteristic of the repayment is that the debtor's payments are broken down into an **interest portion** and a **repayment portion**. Under the interest portion, one understands the interest due at the end of each period on the outstanding principal at the beginning of the period. The repayment portion includes the amount of the repayment at the end of the period in addition to the interest due. The sum of all repayments made over time is

7.2 Repayment Calculation

exactly the credit amount. The sum of the interest to be paid and the repayment made in one period is called the **annuity**.

▶ **Definition (Annuity)** If one denotes with T_t, Z_t and A_t the amount of principal, the amount of interest and the annuity, respectively, due at the end of each period t, then the following basic relationship applies for each period:

$$Z_t + T_t = A_t. \tag{7.8}$$

Depending on the amount and/or the temporal distribution of the principal payments or annuities, different **types of repayment** or **types of debt** are distinguished:

1. **General repayment debt**
   In this type of credit, services and counter-services are carried out in an irregular manner.
2. **Total debt without full interest accumulation**
   In this type of repayment, the entire repayment of the credit amount K_0 is made in a single payment at the end of the term. During the term, only the interest due is paid, but no additional repayment is made.

Example Federal Savings Bond Type A.

3. **Total debt with full interest accumulation**
   In this form of credit, in addition to the total capital K_0 at the end of the term, the accumulated interest is also due (rather compound interest). During the credit period, neither interest nor repayment is made. Therefore, all annuities—with the exception of the last one—are 0.

Example Federal Savings Bond Type B.

4. **Repayment by Installments**
   In this form, the repayment takes place at the end of each period in equal repayment rates. Accordingly, it applies:

$$T_1 = T_2 = \cdots = T_n = T.$$

With a term of n years, the original credit amount K_0 is repayed annually with

$$T = \frac{K_0}{n}.$$

Since the repayment also decreases over time, the annuity payments must also decrease with a constant repayment rate.
5. **Annuity Repayment**
   Characteristic of this form is the **constant annuity** during the term.

Notes

1. It is assumed that all payments are made at interest payment dates and are immediately offset.
2. Interest is calculated using so-called **deferred** interest, which means that the interest amount Z_t at the end of an interest period is based on the outstanding balance K_{t-1} at the beginning of this interest period:

$$Z_t = K_{t-1} \cdot p,$$

where p denotes the interest rate.
3. In addition to interest and principal payments, annuities may also include other components such as fees, commissions, etc. This is not considered.

The so-called standard case of the annuity loan then exists when, under the premise

$$\text{interest period} = \text{payment period}$$

a payment structure as in Fig. 7.2 exists. This means that the performance of the creditor lies in the provision of the loan amount, while the consideration of the borrower lies in the payment of equal annuities A starting one period after the loan disbursement.

With these assumptions, we now first set up a *recursive* form for the outstanding debts after t periods. K_0 denotes the **loan amount** at time $t=0$, i.e. at the beginning of the first period. K_t stands for the **outstanding debts** at the beginning of the subsequent period

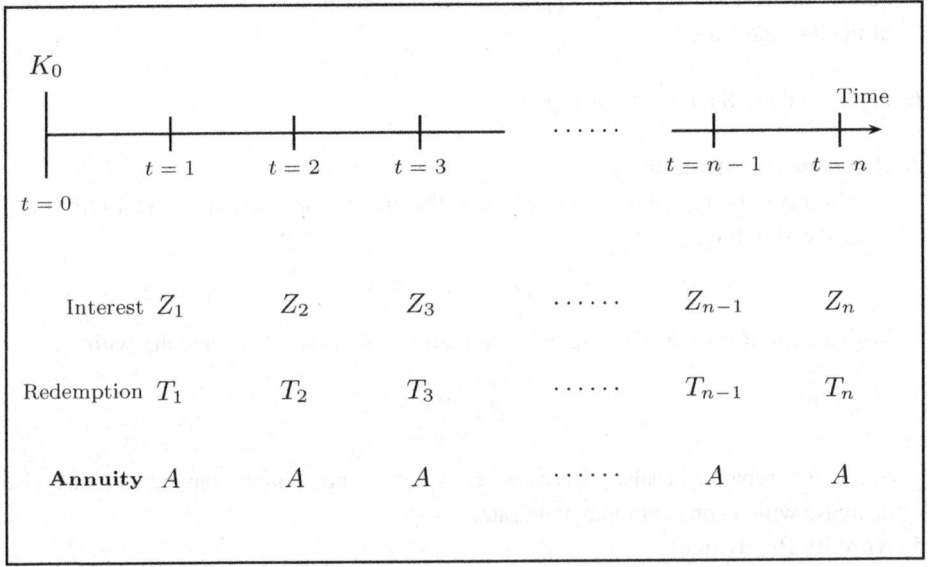

Fig 7.2 Temporal course of an annuity repayment

7.2 Repayment Calculation

$t+1$ The outstanding debts K_t at the end of the period t is determined by the accrual of the outstanding debts K_{t-1} at the beginning of the period minus the annuity paid at the end of the period t. Quantitatively, this means:

Recursive form of the repayment schedule:

$$K_t = K_{t-1}(1+i) - A_t, \quad \text{with } i = \frac{p}{100}. \tag{7.9}$$

Recursive means that the debt in period t is determined from the previous period $t-1$.

The complete course of a loan transaction with all payments, the determination of interest and repayment amounts, and the development of the remaining debt is represented in an **repayment schedule**.

In the following, we consider the very common case that the annuity is constant, i.e.

$$A_t = A = const.$$

This has the consequence that in each period the sum of interest and repayment is the same. The interest payments then decrease from period to period, while the repayment share increases continuously. The repayment schedule then lists for each period:

- the remaining debt

$$K_t = K_{t-1}(1+i) - A$$

- the interest

$$Z_t = K_{t-1} \cdot i$$

- the payment

$$T_t = A - Z_t.$$

Example For a loan of $K_0 = 350.000$ € at an interest rate of $p = 1.5\%$ and an annuity of $A = 20.000$ € in the year in the table. 7.1 repayment schedule.

The repayment schedule provides a detailed view of the entire repayment of a loan. Often one is interested in determining important sizes of an annuity loan directly. Therefore, a closed form for the remaining debt is sought in contrast to the recursive form of Eq. (7.9).

Starting from

$$K_t = K_{t-1}(1+i) - A$$

we get with $q = 1+i$ the form

$$K_t = K_{t-1} \cdot q - A.$$

Tab. 7.1 Repayment schedule

Period	Outstanding debt in €	Interest in €	Repayment in €	Costs in €
0	350.000,00	0	0	0
1	335.250,00	5250,00	14.750,00	20.000
2	320.278,75	5028,75	14.971,25	40.000
3	305.082,93	4804,18	15.195,82	60.000
4	289.659,18	4576,24	15.423,76	80.000
5	274.004,06	4344,89	15.655,11	100.000
6	258.114,12	4110,06	15.889,94	120.000
7	241.985,84	3871,71	16.128,29	140.000
8	225.615,62	3629,79	16.370,21	160.000
9	208.999,86	3384,23	16.615,77	180.000
10	192.134,86	3135,00	16.865,00	200.000
11	175.016,88	2882,02	17.117,98	220.000
12	157.642,13	2625,25	17.374,75	240.000
13	140.006,76	2364,63	17.635,37	260.000
14	122.106,86	2100,10	17.899,90	280.000
15	103.938,47	1831,60	18.168,40	300.000
16	85.497,54	1559,08	18.440,92	320.000
17	66.780,01	1282,46	18.717,54	340.000
18	47.781,71	1001,70	18.998,30	360.000
19	28.498,43	716,73	19.283,27	380.000
20	8925,91	427,48	19.572,52	400.000

The remaining debt can be transformed by successively replacing it with the corresponding expression of the previous period:

$$K_t = K_{t-1} \cdot q - A$$
$$= (K_{t-2} \cdot q - A) \cdot q - A$$
$$= K_{t-2} \cdot q^2 - A \cdot q - A.$$

If you carry this out exactly t times, you will have the following expression for the residual debt:

$$K_t = K_0 \cdot q^t - A \sum_{i=1}^{t} q^{i-1}. \qquad (7.10)$$

Using the geometric series (see Eq. (2.12)), we can give a closed form of the repayment formula:

7.2 Repayment Calculation

Closed form of the Repayment Schedule
For a constant annuity A it holds:

$$K_t = K_0 \cdot q^t - A \cdot \frac{q^t - 1}{q - 1}. \tag{7.11}$$

Example The credit amount is 400,000.00 €, $p = 1.5\%$ p.a. and $A = 12{,}000.00$ € per year. The residual debt K_9 after payment of the 9th annuity then results in:

$$K_9 = K_0 \cdot q^9 - A \frac{q^9 - 1}{q - 1}$$

$$= 400.000 \text{ €} \cdot 1.015^9 - 12.000 \text{ €} \cdot \frac{1.015^9 - 1}{1.015 - 1}$$

$$= 342.622{,}01 \text{ €}.$$

Because of the relationship

$$A = Z_{10} + T_{10} = K_9 \cdot i + T_{10}$$

the repayment T_{10} in the 10th year then results in:

$$T_{10} = 12.000 \text{ €} - 342.644 \text{ €} \cdot 0,015 = 6.860{,}34 \text{ €}.$$

Of particular interest is the **term** $t = n$ of an annuity loan, that is, the question of when the residual debt is zero ($K_n = 0$):

$$K_n = K_0 \cdot q^n - A \cdot \frac{q^n - 1}{q - 1} \stackrel{!}{=} 0.$$

If you solve this equation for n, you get:

$$K_0 q^n (q - 1) - A(q^n - 1) = 0$$
$$q^n (A - K_0(q - 1)) = A.$$

Duration of an Annuity Loan

$$n = \frac{\ln\left(\frac{A}{A - K_0(q - 1)}\right)}{\ln q}. \tag{7.12}$$

With annuity loans, a **percentage rate of repayment** τ of the loan K_0 is often estimated for the first period, i.e. $T_1 = \tau \cdot K_0$. The annuity is in this case given by:

$$A = T_1 + Z_1$$
$$= \tau \cdot K_0 + i \cdot K_0$$
$$= K_0 \cdot (\tau + i).$$

We now ask how the remaining debt K_{m_r} decreases with the duration of the loan. (K_{m_r}: Outstanding debt after m_r payment periods). Let

$$K_{m_r} = r \cdot K_0 \qquad (0 \leq r \leq 1).$$

From Eq. (7.11)

$$K_{m_r} = K_0 \cdot q^{m_r} - A \cdot \frac{q^{m_r} - 1}{q - 1}$$

with $A = K_0(\tau + i)$:

$$r \cdot K_0 = K_0 \cdot q^{m_r} - K_0(\tau + i)\frac{q^{m_r} - 1}{q - 1}. \qquad (7.13)$$

We solve Eq. (7.13) for m_r, set $q = 1 + i$ and obtain:

$$m_r = \frac{1}{\ln(1+i)} \cdot \ln\frac{\tau + i(1-r)}{\tau} = \frac{\ln\left(1 + \frac{i}{\tau}(1-r)\right)}{\ln(1+i)}. \qquad (7.14)$$

This time is independent of the loan amount under the given conditions. For $r = 0$ we get the time after which the debt is fully repaid.

Example A loan is agreed at an interest rate of 1.5% per year, the initial repayment is 2.5% of the loan amount. After how much time is 60% of the loan repaid?

The remaining debt is then 40%, i.e. $K_m = 0.4 \cdot K_0$. For the required time we get from Eq. (7.14):

$$m_{0,4} = \frac{\ln\left(1 + \frac{0.015}{0.025}(1 - 0.4)\right)}{\ln(1 + 0.0125)} = 43.11 \text{ years}.$$

The loan is completely repaid after the time m_0:

$$m_0 = \frac{1}{\ln 1.0125} \cdot \ln\left(\frac{0.025 + 0.015}{0.025}\right) = 61.54 \text{ years}.$$

From this consideration it follows that with annuity debts the repayment only starts slowly at first. In general, the time behaviour of the outstanding debt can be examined by a discussion of the function

$$r(m) = q^m - (p + \tau) \cdot \frac{q^m - 1}{q - 1} = q^m - \left(1 + \frac{\tau}{p}\right) \cdot (q^m - 1).$$

The graph of this function is shown in Fig. 7.3. The values $p = 1.5\%$, $\tau = 2.5\%$ and $q = 1.015$ were used. On the y-axis, the ratio of remaining debt to total debt is plotted, the x-axis corresponds to time.

7.3 Exercises

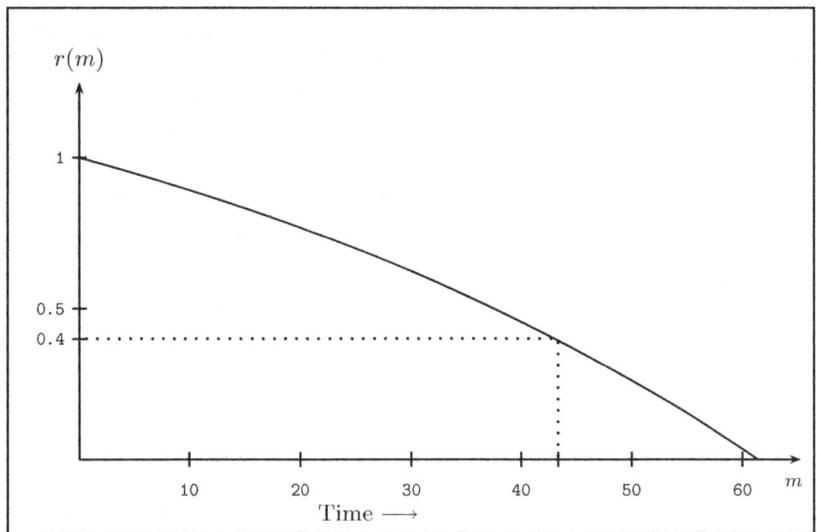

Fig. 7.3 The ratio of the remaining debt to the loan until full repayment for the interest rate $p=1.5\%$ and the repayment rate $\tau=2.5\%$

7.3 Exercises

Short solutions to the following exercises can be found in the appendix.

7.1 A capital K_0 of 5000,– € is deposited on a savings book with an interest rate of 3% from 13.1. to 27.6. What is the final value of the investment?

7.2 A claim falling due on the third of March in the amount of 7200,– € is only paid on twenty-second of September with an amount (incl. Late payment interest) of 7882.50 €. What is the annual interest rate?

7.3 What amount must a saver deposit on fifth of February at 3% p.a. in order to have 10,000,– € at the end of the year?

7.4 How long must a capital of 10,000,– € be invested at 2% p.a. in order to result in an end capital of 11,000,– €? Calculate the term

1. with the linear interest,
2. with the compound interest formula.

7.5 What amount will 8000 € grow to in 10 years at 1.5% compound interest?

7.6 A saver pays 2000 € into an account that is blocked for 7 years. The annual interest rate is 1.5% at the beginning, after three years 1.75%. Calculate the account balance after the block is lifted.

7.7 The interest calculation is not only used in financial mathematics, as is shown in the following task. In what time does a population double at an annual growth rate of 1.3%?

7.8 A debt of 350,000.00 euros is to be interest-bearing and repaid in 20 years with equal annuities at $p = 1.5\%$ per year. The following quantities are to be determined:

- the annuity,
- the repayment at the end of the last year,
- the remaining debt after 10 years,
- the repayment plan,
- the total term if the annuity is given with 14,000 €/year, 16,000 €/year and 18,000 €/year.

7.9 For "house buyers" the loan conditions have been very different in recent years. How long does the repayment of an annuity loan take if 2.5% of the loan amount is agreed as initial repayment and the interest per year is 1.5%? How does this time change if the repayment is 1.5% and the interest is 2.5%?

7.10 An effective annual interest rate of $i_{eff} = 0.025$ is agreed for a loan, the interest surcharge is made monthly. What is the nominal annual interest rate?

7.11

(a) An annuity loan of 350,000 € is repaid under the following conditions: The interest rate is 2%, the annuity 15,000 € per year. How much does the borrower pay to the bank until the debt is repaid?
(b) How does this amount change if the borrower makes a special repayment of 50,000 € after 10 years?

Appendix A

A.1 Solutions to the Test

Solutions to the Test from Sect. 1.6

1.1

$$A \cup B = \{0, \ldots, 10, 12, 14, 16, 18, 20\}$$
$$A \cap B = \{0, 2, 4, 6, 8, 10\}$$
$$A \setminus B = \{1, 3, 5, 7, 9\}$$

1.2

(a) The two statements $A(x)$ and $B(x)$ are equivalent, because
$$x^2 = 16 \iff x = 4 \vee x = -4.$$
Thus, $A(x) \Leftrightarrow B(x)$.

(b) Yes, because:
$$A(x): \quad x^2 - y^2 = 0 \iff (x-y)(x+y) = 0$$
$$\iff x = y \vee x = -y$$
and therefore is $A(x) \Leftrightarrow B(x)$.

(c) No, because
$$x^2 \geq a \iff |x| \geq \sqrt{a} \iff x \geq \sqrt{a} \vee x \leq -\sqrt{a}.$$
For this reason, $A(x) \not\Leftarrow B(x)$.

© The Editor(s) (if applicable) and The Author(s), under exclusive license to Springer-Verlag GmbH, DE, part of Springer Nature 2023
T. Holey and A. Wiedemann, *Analysis and Linear Algebra*,
https://doi.org/10.1007/978-3-662-66247-2

1.3

(a) A necessary condition for winning the lottery is submitting the lottery ticket. A sufficient condition for winning is marking the correct numbers.

(b) No, it applies $B(x) \Longrightarrow A(x)$, but $A(x) \not\Longrightarrow B(x)$.

1.4

(a) $\frac{x+2}{x^2-4} + \frac{1}{x+2} = \frac{2x}{x^2-4}$.

(b) It holds:
$$\frac{3x^n + 2x^{n+2}}{x^{n+1} + 3x^n} = \frac{x^n(3 + 2x^2)}{x^n(x+3)}$$
$$= \frac{3 + 2x^2}{x+3}.$$

This term is defined for $x \neq 0$, $x \neq -3$.

(c) We carry out the following simplification:
$$\sqrt[3]{x^{6n-9}} = \left(x^{6n-9}\right)^{\frac{1}{3}}$$
$$= x^{2n-3}$$
$$= \frac{x^{2n}}{x^3}.$$

This term is defined for all $x \neq 0$.

1.5 Solving equations:

(a)
$$2x - 7 = \tfrac{3}{2}x + \sqrt{3}$$
$$\Longleftrightarrow \quad \tfrac{1}{2}x = \sqrt{3} + 7$$
$$\Longleftrightarrow \quad x = 2(\sqrt{3} + 7) \approx 17,46.$$

(b)
$$\frac{x-3}{2x+6} = 4 \quad (x \neq 3)$$
$$\Longleftrightarrow \quad x - 3 = 4(2x+6)$$
$$\Longleftrightarrow \quad x - 3 = 8x + 24$$
$$\Longleftrightarrow \quad 7x = -27$$
$$\Longleftrightarrow \quad x = -\tfrac{27}{7}.$$

Appendix A

(c) The quadratic equation

$$3x^2 + 2x - 1 = 0$$

has the two solutions

$$x_{1/2} = \frac{-2 \pm \sqrt{4 + 12}}{6} = \frac{-1 \pm 2}{3},$$

so:

$$x_1 = -1, \quad x_2 = \frac{1}{3}.$$

(d) The equation

$$x - 3 = \frac{1}{2 + x}$$

leads to the quadratic equation

$$x^2 - x - 7 = 0$$

with the solutions:

$$x_{1/2} = \frac{1 \pm \sqrt{1 + 28}}{2} = \frac{1 \pm \sqrt{29}}{2}.$$

(e) Squaring the equation

$$\sqrt{x} = 1 - x$$

leads to the quadratic equation

$$x^2 - 3x + 1 = 0.$$

Of the two solutions, only $x = (3 - \sqrt{5})/2$ is the solution of the original equation, since squaring is not an equivalence transformation.

(f)

$$x^5 - 12 = 3$$
$$x^5 = 15$$
$$x = \sqrt[5]{15} \approx 1{,}7187.$$

(g) To solve the equation

$$x^4 + 4x^2 - 8 = 0$$

we substitute: $x^2 = u$ bzw. $x = \pm\sqrt{u}$. Substitution gives the quadratic equation:

$$u^2 + 4u - 8 = 0$$

with the two solutions:
$$u_{1/2} = -2 \pm 2 \cdot \sqrt{3}.$$

Resubstitution gives:
$$x_1 = \sqrt{-2 + 2\sqrt{3}}; \quad x_2 = -\sqrt{-2 + 2\sqrt{3}}.$$

The solution $u_2 = -2 - 2\sqrt{3}$ does not provide (real) solutions for x.

(h)
$$3^x + 12 = 24$$
$$3^x = 12$$
$$x = \log_3 12$$
$$= \frac{\ln 12}{\ln 3}$$
$$\approx 2{,}261.$$

(i)
$$3^x + 3^{x+2} = 110$$
$$3^x + 9 \cdot 3^x = 110$$
$$10 \cdot 3^x = 110$$
$$3^x = 11$$
$$x = \log_3 11$$
$$= \frac{\ln 11}{\ln 3}$$
$$\approx 2{,}18.$$

(j) To solve the equation
$$-2^x + 4 \cdot 2^{2x} = 128$$

we substitute
$$2^x = u, \quad x = \log_2 u, \quad (u > 0), \qquad \text{with } u^2 = (2^x)^2 = 2^{2x}$$

Substituting into the above equation gives the quadratic equation:
$$4u^2 - u - 128 = 0$$

with the two solutions:
$$u_{1/2} = \frac{1 \pm \sqrt{2049}}{8}.$$

Since $u>0$, the solution sought is

$$u_1 = \frac{1+\sqrt{2049}}{2}.$$

So

$$x = \log_2\left(\frac{1+\sqrt{2049}}{8}\right) \approx 2{,}531.$$

(k)

$$\log_2 4x = 15$$
$$4x = 2^{15}$$
$$x = \frac{1}{4}\cdot 2^{15} = 2^{13} = 8192.$$

(l)

$$\lg(x+10) - \lg(2x+5) = 3 \qquad (x>3 \wedge x > \frac{5}{2})$$
$$\lg\frac{x+10}{2x+5} = 3$$
$$\frac{x+10}{2x+5} = 10^3$$
$$x+10 = 2000x + 5000$$
$$x = -\frac{4990}{1999} \approx -2{,}496.$$

(m)

$$\log_2 3x + \log_4 5x = 3$$
$$\log_2 3x + \frac{\log_2 5x}{\log_2 4} = 3$$
$$\log_2 3x + \frac{1}{2}\log_2 5x = 3$$
$$\log_2 3x + \log_2 \sqrt{5x} = 3$$
$$\log_2(3x\sqrt{5x}) = 3$$
$$3x\sqrt{5x} = 2^3$$
$$3\sqrt{5}\cdot x^{3/2} = 8$$
$$x = \left(\frac{8}{3\sqrt{5}}\right)^{2/3}.$$

1.6 The positive real number x is sought so that $1/x$ is 1 less than x. This means that x is the solution of the equation:

$$\frac{1}{x} = x - 1$$

or

$$x^2 - x - 1 = 0.$$

This quadratic equation has the two solutions

$$x_{1/2} = \frac{1 \pm \sqrt{5}}{2}.$$

Since the positive solution is sought, it follows:

$$x_1 = \frac{1 + \sqrt{5}}{2} \approx 1{,}61803\ldots.$$

It follows from the quadratic equation that the square of this number is 1 greater than x. Therefore:

$$x = 1{,}61803\ldots$$
$$\frac{1}{x} = 0{,}61803\ldots$$
$$x^2 = 2{,}61803\ldots.$$

The number $x = 1{,}61803\ldots$ is also called Golden Ratio.

1.7 According to the binomial formula Eq. (1.19) one obtains:

$$(a+b)^4 = \sum_{i=0}^{4} \binom{4}{i} a^i b^{4-i}$$
$$= a^4 + 4a^3 b + 6a^2 b^2 + 4ab^3 + b^4.$$

1.8 A visualization of the relationship

$$\binom{n}{k} = \binom{n-1}{k-1} + \binom{n-1}{k}$$

The relationship between the binomial coefficients is the Pascal triangle. This relationship states that the k-th element of the n-th row is the sum of the elements $k-1$ and k of the previous row.

Appendix A

$$\begin{array}{ccccccccc} & & & & 1 & & & & n=0 & k=0 \\ & & & 1 & & 1 & & & n=1 & k=0,1 \\ & & 1 & & 2 & & 1 & & n=2 & k=0,1,2 \\ & 1 & & 3 & & 3 & & 1 & n=3 & k=0,1,2,3 \\ 1 & & 4 & & 6 & & 4 & & 1 \quad n=4 & k=0,1,2,3,4. \end{array}$$

For the formal proof of the relationship between the binomial coefficients, one uses Eq. (1.20):

$$\binom{n-1}{k-1} + \binom{n-1}{k} = \frac{(n-1)!}{(k-1)!(n-1-k+1)!} + \frac{(n-1)!}{k!(n-1-k)!}$$

$$\stackrel{*}{=} \frac{n!k}{nk!(n-k)!} + \frac{n!(n-k)}{nk!(n-k)!}$$

$$= \frac{n!(k+n-k)}{nk!(n-k)!}$$

$$= \frac{n!n}{nk!(n-k)!}$$

$$= \frac{n!}{k!(n-k)!}$$

$$= \binom{n}{k}.$$

In step * we use the definition of the factorial in the form:

$$(n-1)! = \frac{n!}{n}.$$

1.9

(a)
$$3x + 5 \geq -2x - 3$$
$$5x \geq -8.$$

(b)
$$x^5 > 125$$
$$x > \sqrt[5]{125}.$$

(c)
$$-2x^2 + 3x - 1 < 0$$
$$2x^2 - 3x + 1 > 0.$$

First, the corresponding equation is solved; this leads to the solutions:

$$x_1 = 1, \quad x_2 = \frac{1}{2}.$$

This can be written as the inequality:

$$(x-1) \cdot (x - \frac{1}{2}) > 0.$$

This inequality is only fulfilled if both factors are greater than zero or both are less than zero. The solution set of the inequality results from this to:

$$L = \{x \in \mathbb{R} \mid x > 1 \vee x < \frac{1}{2}\}.$$

1.10 Multiplying the equation

$$x_1 + x_2 = p$$

with x_1, results in

$$x_1^2 + x_1 \cdot x_2 = p \cdot x_1$$

or with $x_1 \cdot x_2 = q$:

$$x_1^2 - p \cdot x_1 + q = 0.$$

In other words, x_1 is a solution of the quadratic equation $x^2 - px + q = 0$, for symmetry reasons the same is true for x_2.

Conversely, if x_1, x_2 are solutions of the quadratic equation

$$x^2 - px + q = 0,$$

then

$$0 = (x - x_1)(x - x_2) = x^2 - (x_1 + x_2)x + x_1 \cdot x_2$$

and

$$x_1 + x_2 = p, \quad x_1 \cdot x_2 = q.$$

A.2 Solutions to the Exercises

The solutions to the exercises are given or briefly outlined here.

Chapter 2

2.1

(a) $D_f^{max} = \{x \in \mathbb{R} \mid x \geq 2 \lor x \leq 1\}$,
(b) $D_f^{max} = \{x \in \mathbb{R} \mid -1 < x < +1 \land x \neq 0\}$.

2.2

$(f \circ f)(x) = x^4 - 10x^2 + 20, (f \circ g)(x) = \dfrac{25x^2}{(x^2 - 2)^2} - 5$,

$(g \circ f)(x) = \dfrac{5x^2 - 25}{(x^2 - 5)^2 - 2}$.

$f(g(1)) = 20, \qquad g(f(1)) = -\dfrac{10}{7}$.

2.3
$A = 2, b = 3\pi, x = 1/2$, Zeros: $x = \frac{1}{6} + \frac{n}{3}, n \in \mathbb{Z}$.

2.4
$W_f = \{y \in \mathbb{R} \mid -3 \leq y \leq +3\} \subset \mathbb{R}$

2.5

1. $D_{max} = \mathbb{R}$

2. $W_f = \begin{cases} \{y \in \mathbb{R} \mid y \geq c - \frac{b^2}{4a}\}, & a > 0, \\ \{y \in \mathbb{R} \mid y \leq c - \frac{b^2}{4a}\}, & a < 0. \end{cases}$

3. $\dfrac{b^2}{4a} + 1 \leq c$.

2.6

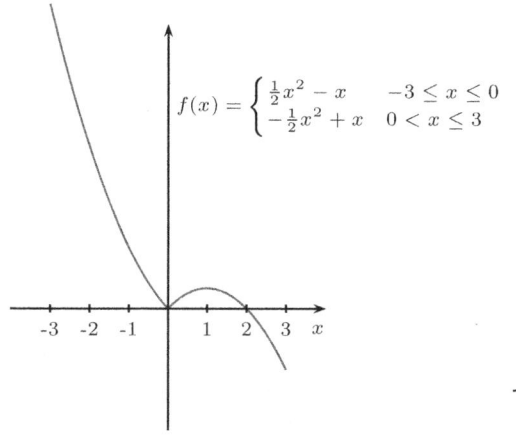

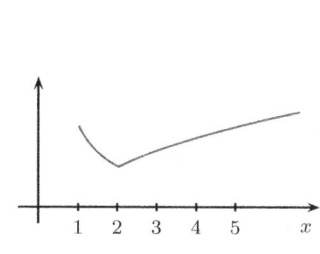

2.7 $k \approx 0{,}025$, $x_D \approx 28{,}07$ years.

2.8

(a) Value range: $W_f = \{y \in \mathbb{R} \mid -2 \leq y \leq +8\}$,
Inverse function: $y = g(x) = 2x - 6$.
(b) Value range: $W_f = \{y \in \mathbb{R} \mid y \geq 1\}$,
Inverse function: $g(x) = +\frac{1}{2}\sqrt{x-1}$.

2.9 (a) bounded above, (b) bounded above, (c) bounded below, (d) bounded.

2.10 Injective, not surjective, not bijective.

2.11 (a) Not injective, not surjective, not bijective. (b) Not injective, surjective, not bijective. (c) Injective, surjective and bijective. (d) Not injective, surjective, not bijective.

2.12 $f(x)$ is continuous, if $a = 3/e$.

2.13 (a) $D_f = \mathbb{R}, P_0 = (\ln 3^{1/3}, 0)$. (b) $D_f = \mathbb{R}$, no zero. (c) $D_f = \mathbb{R}, P_0 = (0,0)$. (d) $D_f = \mathbb{R}, P_{0,1} = (\pm 2, 0)$. (e) $D_f = \mathbb{R} \setminus \{-3\}$, no zeros.

2.14

(a) Domain: $D_f = \mathbb{R}$, inverse image exists for $x \geq 0$.
Zero points: $P_0 = (0,0)$
Range: $W_f = \{y \in \mathbb{R} \mid y \geq 0\}$.
Inverse function: $g(x) = +\sqrt{\exp\{2x\} - 1}$.
(b) Definition area: $D_g = \mathbb{R}_+$.
Zero points: $l = 2$.
Value range: $W_g = \mathbb{R}$.
Inverse function: $l(y) = 2 \cdot e^y$.
(c) Domain: $D_f = \mathbb{R}_+$.
Zero points: $x_0 = \frac{-1+\sqrt{5}}{2}$.
Inverse function: $g(x) = \frac{1}{2}(\sqrt{1 + 4e^x} - 1)$.
(d) Domain: $D_h = \{b \in \mathbb{R} \mid b > 1\}$.
Zero points: $b_0 = \sqrt{\frac{1+\sqrt{5}}{2}}$.
Inverse function: $g(x) = +\sqrt{\frac{1}{2}(\sqrt{1 + 4e^{2x}} + 1)}$.

2.15 $f(x)$ even, then $f(x)-f(-x)=0$.

$$f(x)-f(-x) = \frac{x}{e^x-1} + \frac{x}{2} + \frac{x}{e^{-x}-1} + \frac{x}{2}$$
$$= x\left(\frac{e^{-x}-1+e^x-1}{(e^x-1)(e^{-x}-1)}+1\right)$$
$$= x\left(\frac{e^{-x}+e^x-2}{-e^x-e^{-x}+2}+1\right)$$
$$= x(-1+1) = 0.$$

2.16 Since $f(x) = \sin x$ is defined on every interval

$$-\frac{\pi}{2}+n\cdot\pi \le x \le \frac{\pi}{2}+n\cdot\pi, \quad n \in \mathbb{Z}$$

there are two sets of infinitely many inverse functions

$$y = 2n\pi + \arcsin x \quad \text{und } y = (2n+1)\pi - \arcsin x, \quad n \in \mathbb{Z}.$$

Therefore, we obtain the two sets of inverse functions

$$x = \frac{1}{b}\left(\arcsin\left(\frac{d}{A}\right)+2n\pi\right) - c, \quad n \in \mathbb{Z},$$
$$x = \frac{1}{b}\left((2n+1)\cdot\pi - \arcsin\left(\frac{d}{A}\right)\right) - c, \quad n \in \mathbb{Z}.$$

2.17 $f(x) = \dfrac{U_{max}}{2}\left(\sin\left[\dfrac{\pi}{6}(x-3)\right]+1\right).$

2.18 If you apply the quotient criterion, you get the case that the quotients converge to 1. The criterion makes no statement about this. The harmonic series diverges, the $1/n^2$ series converges.

2.19 (a) $R=10$, (b) $R=1$, (c) $R=\infty$, (d) $R=e$.

2.20 (a) 0, (b) 0, (c), (d) does not exist.

2.21 $c=10$; $K_F=1000$.

2.22 (a) $f_\infty=32$, (b) $t_{50\%}=2{,}77$, $t_{70\%}=4{,}46$, $t_{90\%}=7{,}16$.

2.23 $p(x_N) = \frac{x_0}{c} - \frac{1}{c}x_N(p)$, monotonically decreasing.

2.24 $p_{1/2} = \dfrac{-c \pm \sqrt{c^2-4b(a_0-x_0)}}{2b}$, due to $p>0$ $a_0<x_0$ must be fulfilled in order for a market equilibrium to occur.

2.25 (a) $x_1=-1, x_2=-2, x_3=-3$, b) $x_1=2, x_2=x_3=-5, x_4=3$.

2.26 (a) 2, (b) -2, (c) 2

Chapter 3

3.1 $\lim_{x \to 1} \frac{e^x-1}{\sqrt{x-1}} = \infty$ (de L'Hospital not applicable), $\lim_{x \to 1} \frac{\ln x}{x-1} = 1$.

3.2

(a) $f'(x)=-3$, applying the limit process:

$$f'(x) = \lim_{\Delta x \to 0} \frac{-3(x+\Delta x) + 8 + 3x - 8}{\Delta x} = -3.$$

(b) $f'(x)=2x$, applying the limit method:

$$f'(x) = \lim_{\Delta x \to 0} \frac{(x+\Delta x)^2 + a^2 - (x^2 + a^2)}{\Delta x} = 2x.$$

(c) $f'(x)=2a(ax+b)$, applying the limit value method:

$$f'(x) = \lim_{\Delta x \to 0} \frac{(a(x+\Delta x)+b)^2 - (ax+b)^2}{\Delta x}$$
$$= \lim_{\Delta x \to 0} (2a(ax+b) + a^2 \Delta x)$$
$$= 2a(ax+b).$$

3.3 (a) $f'(x) = \frac{e^x(x-1)}{x^2}$, (b) $f'(x) = \frac{4x+3}{2x^2+3x+5}$, (c) $f'(x) = \frac{1}{5}x^{-4/5}$, (d) $f'(x) = 1 + \ln x$.

3.4

$$f'(x) = \lim_{\Delta x \to 0} \frac{\ln(x+\Delta x) - \ln x}{\Delta x}$$
$$= \lim_{\Delta x \to 0} \frac{1}{\Delta x} \ln \frac{x+\Delta x}{x}$$
$$= \lim_{\Delta x \to 0} \ln\left(1 + \frac{\Delta x}{x}\right)^{1/\Delta x}$$
$$= \lim_{\Delta x \to 0} \frac{1}{x} \ln\left(1 + \frac{\Delta x}{x}\right)^{x/\Delta x}.$$

With $h = \Delta x/x$ ($\Delta x \to 0$ means also $h \to 0$):

$$f'(x) = \lim_{h \to 0} \frac{1}{x} \ln(1+h)^{1/h} = \frac{1}{x} \lim_{h \to 0} \ln(1+h)^{1/h}$$
$$= \frac{1}{x} \ln e = \frac{1}{x}.$$

Appendix A

3.5 (a) continuous and differentiable, (b) continuous, not differentiable.

3.6 The initial solution $x_1 = 1$ yields:

$$x_2 = x_1 - \frac{h(x_1)}{h'(x_1)} = 1 - \frac{e^{-1} - 1}{-e^{-1} - 1} \approx 0{,}538.$$

3.7 $t_1(x) = \frac{1}{20}(x+7), t_2(x) = \frac{1}{5}(x+7).$

3.8 With K_l costs of laying per km on land, K_w costs of laying per km in water, the length-dependent cost function for laying is:

$$K = K_w \cdot s_w + K_l \cdot s_l.$$

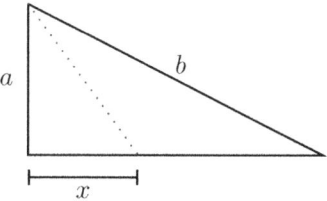

The onshore section is $s_l = \sqrt{b^2 - a^2} - x$, which is in the water $s_w = \sqrt{a^2 + x^2}$. First derivative with respect to x:

$$K'(x) = \frac{K_w \cdot x}{\sqrt{a^2 + x^2}} - K_l \stackrel{!}{=} 0, \quad x_{min} = \frac{a}{\sqrt{\left(\frac{K_w}{K_l}\right)^2 - 1}}.$$

3.9

(a) $f(x) = x^3 - \frac{x^2}{2} + x - 1, f'(x) = 3x^2 - x + 1, f''(x) = 6x - 1.$
(b) $f(x) = e^{-x^2}, f'(x) = -2xe^{-x^2} = -2x \cdot f(x), f''(x) = (4x^2 - 2)f(x).$
(c) $f(x) = 2\sin x + 5\cos x, f'(x) = 2\cos x - 5\sin x,$
$$f''(x) = -2\sin x - 5\cos x = -f(x)$$

3.10 $a = \frac{1}{2}, b = 1, c = 3/2.$

3.11 $b^2 < 3ac.$

3.12

(a) Zeros: $x_1 = 0, x_{2/3} = t^2 \pm \sqrt{t^4 + t}$.
(b) Points with horizontal tangent: $x_{4/5} = \frac{2t^2 \pm \sqrt{4t^4 + 3t}}{3}$, therefore 2 solutions for $t > 0, 1$ solution for $t = 0$. The character of the points with horizontal tangent results from the investigation of the convex and concave areas and the behavior of the function for $|x| \to \infty$.
(c) Symmetry: $f_0(-x) = -f_0(x)$, i. e. for $t = 0$ symmetry to the origin. For $t \neq 0$ no symmetry recognizable.
(d) Convex and concave areas:

$$\text{For } x < \frac{2}{3}t^2 : \quad f_t''(x) > 0 \quad \text{concave from above}$$

$$\text{For } x > \frac{2}{3}t^2 : \quad f_t''(x) < 0 \quad \text{convex from above}$$

(e) Asymptotics: $\lim_{x \to \infty} f_t(x) = -\infty \quad \lim_{x \to -\infty} f_t(x) = +\infty$.

3.13 Continuity of $f(x)$ in $x_0 = 2$ leads to: $4a - 4 = 2b + 1$.
Differentiability of $f(x)$ in $x_0 = 2$ leads to: $4a - 2 = b$.
Thus, the function $f(x)$ is continuous and differentiable, if $b = -3$ and $a = -\frac{1}{4}$.

3.14 $f(x) = ax^4 + bx^3 + cx^2 + dx + e$ with $a = 1/8$, $b = d = 0$, $c = 3/4$, $e = 5/8$. This function has four real zeros: $x_1 = 1, x_2 = +\sqrt{5}, x_3 = -1, x_4 = -\sqrt{5}$, und $f(x)$ has a local maximum at $P_1 = (0, 5/8)$ and in $P_{2/3} = (\pm\sqrt{3}, -1/2)$ has two local minima.

3.15 The Maclaurin series of $f(x) = \dfrac{1}{1-x}$ is identical to the geometric series

$$f(x) = \frac{1}{1-x} = 1 + x + x^2 + x^3 + \cdots = \sum_{n=0}^{\infty} x^n.$$

3.16 $\ln x = (x-1) - \dfrac{(x-1)^2}{2} + \dfrac{(x-1)^3}{3} \mp$.

3.17 In both cases, we get:

$$f(x) = x + x^2 + \frac{x^3}{3} - \frac{x^5}{30} - \frac{x^6}{90} \mp \cdots$$

3.18 $x_1 = -2, x_2 = 1$.

3.19 Price elasticity: $\epsilon_{x,p}(p) = -\frac{p}{10}$, demand increases by 50%, $p = 100$.

3.20 Products with elastic price elasticity: replaceable goods, consumer goods, products with inelastic price elasticity: food, energy.

3.21 Unit costs: $k(x) = \dfrac{K(x)}{x}$, with $k'(x)=0$, results in $K'(x) = \dfrac{K(x)}{x}$. With $K(x)=10+2x^2$ the marginal costs $K'(x)=4x$. Intersection at $x=\sqrt{5}$.

3.22 $K_F = 625$.

3.23 With $E(p)=x_0\cdot p - c\cdot p^2$ and $E'(p_{max})=0, E(p_{max})=100.000$ results $E(p)=2000p - 10p^2$.

3.24 (a) $[2, \dfrac{3}{2}+\sqrt{105}/2]$, (b) $G(10/6+\sqrt{79/9}) \approx 43,28$, (c) $p=14$ €.

3.25 $k = \dfrac{1}{2}, a = e^{1/2}$.

3.26 $p_v = \dfrac{6}{e^{7/6}}$.

Chapter 4

4.1 (a) $F(x)=x^3+c$, (b) $F(x) = \tfrac{2}{3}x\sqrt{2x}+c$ (c) $F(x) = \tfrac{1}{6}(2x-1)^3 + c$, (d) $F(x) = -\tfrac{1}{a}e^{-ax}+c$, (e) $F(x) = -2\sqrt{1-x}+c$, (f) $F(x) = -\cos x + c$.

4.2

$$\int \ln x\,dx = \int 1\cdot \ln x\,dx = x\cdot \ln x - \int x\cdot \frac{1}{x}dx = x\cdot \ln x - x.$$

$$\int_{1/e}^{e} \ln x\,dx = \frac{2}{e}.$$

4.3 $c = \dfrac{1}{b-a}$,

$$F(x) = \begin{cases} 0 & \text{for } x \leq a \\ \int_a^x c\,dx = c(x-a) & \text{for } a \leq x \leq b, \\ c\cdot(b-a) & \text{for } x > b \end{cases}$$

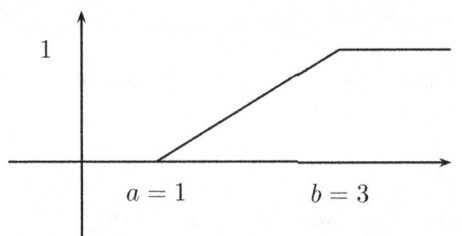

$$\int_{-\infty}^{+\infty} x \cdot f(x)dx = \frac{a+b}{2}. \quad \text{for} \quad c = 1/(b-a).$$

4.4 $A = 11/6$.

4.5 $I = e^2 - 2$.

4.6 $\lim_{b \to \infty} \int_0^b e^{-ax} dx = \frac{1}{a}$.

4.7 $A = \frac{1}{2}$.

4.8 $x_0 = 3$, $K = 18$, $P = 9$.

4.9 Two-fold partial integration yields: $I = -\dfrac{e^{-x}}{2}(\cos x + \sin x)$.

4.10 $I_1 = -\frac{1}{2}e^{-x^2}$, $I_2 = \frac{1}{2}$.

4.11 $I = \frac{1}{4}\ln(x^4 - 5) + c$.

4.12

(a) $t_m = 2\sqrt{3}$ is the time at which the capital flow is at its maximum. Capital flow: $f(t_m) = 48 \cdot \sqrt{3}$.
(b) $f(t) = 0 \Leftrightarrow -t^3 + 36t = 0 \Rightarrow t_1 = 0$, $t_2 = +6$, the solution $t = -6$ is not meaningful. Maximum capital:

$$\int_0^6 f(t)\, dt = -\frac{1}{4}t^4 + 18t^2 \Big|_0^6 = 324.$$

This makes the total capital $K = K_0 + \Delta K = 900$. (Expenses in thousands of Euros)
(c) $\bar{t} = \sqrt{96} \approx 9{,}8$.

4.13 Maximum: $A = A_0 +$ Zugänge $-$ Abgänge $= 10 + 1{,}115 - 0{,}783 = 10{,}331$ (in Hundred thousand).

4.14 $I_1 = \dfrac{1}{4}, I_2 = -\dfrac{16}{3}, I_3 = \dfrac{65}{27}, I_4 = \dfrac{11}{3}$.

4.15 $\int_0^a \dfrac{2x}{1-x^2} dx = -\ln(1-a^2)$. exists for $a<1$.

Chapter 5

5.1 **r**: Raw materials, **z**: Intermediate products, **e**: End products.

$$\mathbf{A} = \begin{pmatrix} 1 & 2 & 2 \\ 1 & 2 & 1 \end{pmatrix} \qquad \mathbf{B} = \begin{pmatrix} 2 & 2 \\ 1 & 1 \\ 1 & 2 \end{pmatrix}.$$

Resource consumption:

$$\mathbf{r} = \mathbf{A} \cdot \mathbf{B} \cdot \mathbf{e}$$

With $r_1 = 124$, $r_2 = 98$ follows $e_1 = 10$, $e_2 = 8$.

5.2 (a) $\mathbf{A} \cdot \mathbf{B}$ does not exist, (b) $\mathbf{B} \cdot \mathbf{A} = \begin{pmatrix} -2 & 11 & 4 \\ -17 & 2 & -3 \end{pmatrix}$,

(c) $\mathbf{A}^\top \cdot \mathbf{B}$ does not exist, (d) $\mathbf{A} \cdot \mathbf{B}^\top = \begin{pmatrix} 9 & 11 \\ 5 & -4 \\ -6 & 4 \end{pmatrix}$,

(e) $\mathbf{A}^2 = \begin{pmatrix} -11 & 11 & 1 \\ 4 & 8 & -3 \\ -4 & -5 & -13 \end{pmatrix}$, (f) $\mathbf{B}^2$ does not exist, (g) $(\mathbf{B}^\top)^2$ does not exist.

5.3 (a) $\mathbf{A} \cdot \mathbf{B} = \begin{pmatrix} 4 & 45 \\ -2 & 1 \\ -2 & 34 \end{pmatrix}$, (b) $\mathbf{A} \cdot \mathbf{B}$ does not exist, (c) $\mathbf{A} \cdot \mathbf{B} = \begin{pmatrix} -4 & 2 \\ 0 & -1 \end{pmatrix}$,

(d) $\mathbf{A} \cdot \mathbf{B} = \begin{pmatrix} 88 & -4 \\ 2 & 5 \end{pmatrix}$.

5.4 (a) $\mathbf{B}^\top (\mathbf{A} \cdot \mathbf{B}^\top)^{-1} = \mathbf{B}^\top ((\mathbf{B}^\top)^{-1} \cdot \mathbf{A}^{-1}) = (\mathbf{B}^\top (\mathbf{B}^\top)^{-1}) \cdot \mathbf{A}^{-1} = \mathbf{A}^{-1}$.(b)

$$\mathbf{A}^\top (\mathbf{A} \cdot \mathbf{B})^\top (\mathbf{A}^{-1})^\top = \mathbf{A}^\top (\mathbf{B}^\top \cdot \mathbf{A}^\top)(\mathbf{A}^{-1})^\top = \mathbf{A}^\top (\mathbf{B}^\top \cdot \mathbf{A}^\top)(\mathbf{A}^\top)^{-1}$$
$$= \mathbf{A}^\top \mathbf{B}^\top = (\mathbf{B}\mathbf{A})^\top.$$

5.5 (a) Set $(\mathbf{A}^{-1})^{-1} = \mathbf{M}$. Thus follows:

$$\begin{aligned} (\mathbf{A}^{-1})^{-1} &= \mathbf{M} \\ \Longleftrightarrow \quad (\mathbf{A}^{-1})(\mathbf{A}^{-1})^{-1} &= \mathbf{A}^{-1}\mathbf{M} \\ \Longleftrightarrow \quad \mathbf{E} &= \mathbf{A}^{-1}\mathbf{M} \\ \Longleftrightarrow \quad \mathbf{A} &= \mathbf{M}. \end{aligned}$$

(b)
$$(\mathbf{A}^{-1})^\mathsf{T} = (\mathbf{A}^\mathsf{T})^{-1}$$
$$\iff (\mathbf{A}^\mathsf{T})(\mathbf{A}^{-1})^\mathsf{T} = \mathbf{A}^\mathsf{T}(\mathbf{A}^\mathsf{T})^{-1}$$
$$\iff (\mathbf{A}^\mathsf{T})(\mathbf{A}^{-1})^\mathsf{T} = \mathbf{E}$$
$$\iff (\mathbf{A}^{-1}\mathbf{A})^\mathsf{T} = \mathbf{E}$$
$$\iff (\mathbf{E})^\mathsf{T} = \mathbf{E}$$
$$\iff \mathbf{E} = \mathbf{E}.$$

(c)
$$(\mathbf{A} \cdot \mathbf{B})^{-1} = \mathbf{B}^{-1} \cdot \mathbf{A}^{-1}$$
$$\iff (\mathbf{AB})(\mathbf{A} \cdot \mathbf{B})^{-1} = (\mathbf{A} \cdot \mathbf{B})(\mathbf{B}^{-1} \cdot \mathbf{A}^{-1})$$
$$\iff \mathbf{E} = \mathbf{A} \cdot (\mathbf{BB}^{-1}) \cdot \mathbf{A}^{-1}$$
$$\iff \mathbf{E} = \mathbf{A} \cdot \mathbf{A}^{-1}$$
$$\iff \mathbf{E} = \mathbf{E}.$$

5.6
$$\mathbf{A}^{-1} = \frac{1}{ad - bc} \cdot \begin{pmatrix} d & -b \\ -c & a \end{pmatrix}$$

$\mathbf{A}^{-1}$ exists only if det $\mathbf{A} \neq 0$, also $ad \neq bc$.

5.7 The pivot method leads in the second step to

x_1	x_2	x_3	x_4	
1	0	-3	4	-2
0	1	4	-5	5
0	0	0	0	-4

so the system of linear equations is not solvable.

5.8 With the pivot method results after two steps:

x_1	x_2	x_3	
1	0	2	$3a - b$
0	1	$-\frac{5}{2}$	$\frac{b-2a}{2}$
0	0	0	$c - 5a + 2b$

The system of linear equations is solvable, if $c-5a+2b=0$, a unique solution does not exist in any case. The general solution is: $x_1 = 3a - b - 2x_3, x_2 = \frac{b-2a}{2} + \frac{5}{2}x_3, x_3 = x_3$ with $x_3 \in \mathbb{R}$.

5.9 (a)
$$\mathbf{A}^{-1} = \begin{pmatrix} \frac{1}{2} & -\frac{3}{4} & 2 \\ 0 & \frac{1}{2} & -1 \\ -\frac{1}{2} & \frac{1}{4} & 0 \end{pmatrix}$$

(b) $\mathbf{A}^{-1}$ does not exist.

5.10 $\lambda_{1/2} = -2, \lambda_3 = 7$.

5.11 Consider the two 2×2 matrices

$$\mathbf{A} = \begin{pmatrix} a_1 & a_2 \\ a_3 & a_4 \end{pmatrix}, \quad \mathbf{B} = \begin{pmatrix} b_1 & b_2 \\ b_3 & b_4 \end{pmatrix}.$$

Then:
$$\det \mathbf{A} = a_1 a_4 - a_2 a_3, \quad \det \mathbf{B} = b_1 b_4 - b_2 b_3.$$

Therefore:
$$\det \mathbf{A} \cdot \det \mathbf{B} = a_1 a_4 b_1 b_4 - a_1 a_4 b_2 b_3 - a_2 a_3 b_1 b_4 + a_2 a_3 b_2 b_3.$$

On the other hand:
$$\mathbf{AB} = \begin{pmatrix} a_1 b_1 + a_2 b_3 & a_1 b_2 + a_2 b_4 \\ a_3 b_1 + a_4 b_3 & a_3 b_2 + a_4 b_4 \end{pmatrix}$$

and thus
$$\det(\mathbf{A} \cdot \mathbf{B}) = (a_1 b_1 + a_2 b_3)(a_3 b_2 + a_4 b_4) - (a_1 b_2 + a_2 b_4)(a_3 b_1 + a_4 b_3)$$
$$= a_1 a_4 b_1 b_4 - a_1 a_4 b_2 b_3 - a_2 a_3 b_1 b_4 + a_2 a_3 b_2 b_3$$
$$= \det \mathbf{A} \cdot \det \mathbf{B}.$$

5.12 Set: x the number of apples, y the number of bananas, z the number of pineapples. To determine if the units $A = 16, B = 20, C = 18$ can be obtained by a mixture of fruit, the following system of equations must be solved:

$$2x + 3y + z = 16$$
$$2x + 2y + 4z = 20$$
$$2x + y + 3z = 18$$

The solution of this equation system is:
$$x = 6,5, \quad y = 0,5, \quad z = 1,5.$$

5.13
$$\mathbf{K} = \mathbf{P} \cdot \mathbf{B} = \begin{pmatrix} 1150 & 650 & 420 & 1800 \\ 1080 & 580 & 400 & 1800 \\ 1050 & 710 & 420 & 1500 \end{pmatrix}.$$

K_{ij}: Cost in the delivery of the three products from supplier i to the factory j.

5.14 The production units fulfill the: system of linear equations
$$\mathbf{x} = \mathbf{A} \cdot \mathbf{x} + \mathbf{b}.$$

The vector $\mathbf{b}$ describes the deliverable quantities. For $\mathbf{b}$ it follows:
$$\mathbf{b} = (\mathbf{E} - \mathbf{A}) \cdot \mathbf{x}.$$

With the given values, one obtains:
$$\mathbf{b} = \begin{pmatrix} 0 \\ 80 \\ 210 \end{pmatrix}.$$

5.15 (a) The Leontief inverse $(\mathbf{E} - \mathbf{A})^{-1}$ must exist and its elements must be positive or zero. For $a < \frac{4}{5}$ these properties are fulfilled. (b) For the given values one obtains:
$$\mathbf{x} = (\mathbf{E} - \mathbf{A})^{-1} \cdot \mathbf{b} = \begin{pmatrix} 125 \\ 90 \end{pmatrix}.$$

5.16 The system of linear equations is:
$$x_5 = 5$$
$$x_4 = 3x_5 + 6$$
$$x_3 = 2x_5$$
$$x_2 = 3x_3 + 3x_4 + 7x_5$$
$$x_1 = 2x_3 + x_5$$

with the solution:
$$x_5 = 5;\ x_4 = 21;\ x_3 = 10;\ x_2 = 128;\ x_1 = 25.$$

5.17 There exist the products $\mathbf{A} \cdot \mathbf{B},\ \mathbf{A}^T \cdot \mathbf{B},\ \mathbf{B}^T \cdot \mathbf{A},\ \mathbf{B}^T \cdot \mathbf{B}^T$

Chapter 6

6.1 (a) $\frac{\partial f}{\partial x_1} = 3x_1^2 + e^{x_2},\ \frac{\partial f}{\partial x_2} = x_1 e^{x_2} + 2x_2 x_3,\ \frac{\partial f}{\partial x_3} = x_2^2,$

(b) $\frac{\partial f}{\partial x_1} = \ln(x_2 x_3) - \frac{1}{x_1 + x_2},\ \frac{\partial f}{\partial x_2} = \frac{x_1 \cdot x_3}{x_2} - \frac{1}{x_1 + x_2},\ \frac{\partial f}{\partial x_3} = \frac{x_1 \cdot x_2}{x_3}.$

Appendix A

6.2

$$\mathbf{grad}\, f(x, y) = \begin{pmatrix} -2(x-1) \\ -2(y-\frac{1}{2}) \end{pmatrix}$$

The contour map is constructed from the requirement $f(x, y) = c$, or $(x-1)^2 + (y-\frac{1}{2})^2 = 16 - c\, (c \leq 16)$. The contour lines are concentric circles around the point $(x, y) = (1, \frac{1}{2})$ with the radius $\sqrt{16-c}$.

$$\mathbf{grad}\, f(0,0) = \begin{pmatrix} 2 \\ 1 \end{pmatrix}, \quad \mathbf{grad}\, f(1,0) = \begin{pmatrix} 0 \\ 1 \end{pmatrix}.$$

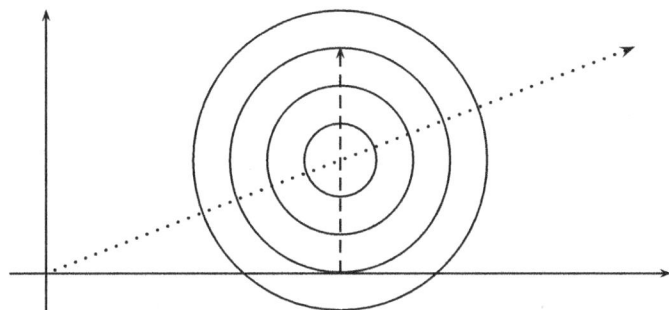

6.3 The output function is: $x(r_1, r_2, r_3) = 3r_1^2 \cdot \sqrt{r_2} + 5r_1 r_2 r_3^{\frac{1}{3}}$. The total differential is:

$$dx = \frac{\partial x}{\partial r_1} dr_1 + \frac{\partial x}{\partial r_2} dr_2 + \frac{\partial x}{\partial r_3} dr_3$$

$$= (6r_1 \sqrt{r_2} + 5r_2 r_3^{1/3}) dr_1 + \left(\frac{3}{2} \frac{r_1^2}{\sqrt{r_2}} + 5r_1 r_3^{1/3}\right) dr_2 + \left(\frac{5}{3} r_1 r_2 r_3^{-2/3}\right) dr_3.$$

At the point $r_1 = r_2 = r_3 = 1$, the following results:

$$dx\bigg|_{(1,1,1)} = 11\, dr_1 + \frac{13}{2} dr_2 + \frac{5}{3} dr_3$$

For the given change, the value is: $dx \approx 2.08$.

6.4 (a) The condition on the production function is:

$$x(r_1, r_2) = 2\sqrt{r_1} \cdot \sqrt[3]{r_2} \stackrel{!}{=} 216.$$

With $r_2 = 27$, the value $r_1 = 1296$ results.

(b) With $dx(r_1, r_2) = 0$, it follows that:

$$\frac{dr_2}{dr_1} = -\frac{\frac{\partial x}{\partial r_1}}{\frac{\partial x}{\partial r_2}} = -\frac{3}{2} \frac{r_2}{r_1}.$$

or $dr_2 = -\frac{3}{2}\frac{r_2}{r_1}dr_1$. With $r_1 = 1296$, $r_2 = 27$ this results in an increase of 0.031 units of the 2nd factor to compensate for the reduction of the 1st factor by one unit.

6.5 The Hessian matrix for functions of two variables, $f(x, y)$, is:

$$\mathbf{H} = \begin{pmatrix} \frac{\partial^2 f}{\partial x^2} & \frac{\partial^2 f}{\partial x \partial y} \\ \frac{\partial^2 f}{\partial x \partial y} & \frac{\partial^2 f}{\partial y^2} \end{pmatrix} = \begin{pmatrix} a & c \\ c & b \end{pmatrix}.$$

The eigenvalues of the Hessian matrix, $\lambda_{1/2}$ are calculated from the condition:

$$\det \begin{pmatrix} a - \lambda & c \\ c & b - \lambda \end{pmatrix} = 0$$

This provides a quadratic equation in λ with the two solutions

$$\lambda_{1/2} = \frac{a + b \pm \sqrt{(a-b)^2 + 4c^2}}{2}.$$

The condition that both eigenvalues have the same sign ($\lambda_1 \cdot \lambda_2 > 0$) leads to $ab - c^2 > 0$ or

$$\frac{\partial^2 f}{\partial x^2} \cdot \frac{\partial^2 f}{\partial y^2} - \left(\frac{\partial^2 f}{\partial x \partial y}\right) > 0.$$

For $a < 0$ and $b < 0$ is:

$$\lambda_1 = \frac{a + b - \sqrt{(a-b)^2 + 4c^2}}{2} < 0$$

with $ab - c^2 > 0$ also $\lambda_2 < 0$, as shown above.

For $a > 0$ and $b > 0$ is:

$$\lambda_1 = \frac{a + b + \sqrt{(a-b)^2 + 4c^2}}{2} > 0$$

with $ab - c^2 > 0$ also $\lambda_2 > 0$, as shown above.

6.6

(a) The Hesse matrix is:

$$H = \begin{pmatrix} -4 & 2 \\ 2 & -12 \end{pmatrix}$$

with eigenvalues

$$\lambda_{1/2} = \frac{-16 \pm \sqrt{88}}{2} < 0.$$

Therefore, $f(x, y)$ is concave from above.

(b)
$$\mathbf{grad} f(x, y) = \begin{pmatrix} 4 - 4x + 2y \\ 2 - 12y + 2x \end{pmatrix}$$

So:
$$| \mathbf{grad} f(0,0) | = \left| \begin{pmatrix} 4 \\ 2 \end{pmatrix} \right| = \sqrt{20} > 0{,}5$$

1. iteration step:

$$f(\mathbf{x}^{(0)} + \mu^{(0)} \cdot \mathbf{grad} f(\mathbf{x}^{(0)})) = f \begin{pmatrix} 4\mu^{(0)} \\ 2\mu^{(0)} \end{pmatrix} = 20\mu^{(0)} - 40(\mu^{(0)})^2$$

$f(\mu^{(0)})$ gets maximum for $\mu^{(0)} = \frac{1}{4}$.
This
$$\mathbf{x}^{(1)} = \begin{pmatrix} 0 \\ 0 \end{pmatrix} + \frac{1}{4} \begin{pmatrix} 4 \\ 2 \end{pmatrix} = \begin{pmatrix} 1 \\ \frac{1}{2} \end{pmatrix}.$$

and
$$| \mathbf{grad} f \begin{pmatrix} 1 \\ \frac{1}{2} \end{pmatrix} | = \left| \begin{pmatrix} 1 \\ -2 \end{pmatrix} \right| = \sqrt{5} > 0{,}5$$

2. iteration step:

$$f(\mathbf{x}^{(1)} + \mu^{(1)} \cdot \mathbf{grad} f(\mathbf{x}^{(1)})) = f \begin{pmatrix} 1 + \mu^{(1)} \\ \frac{1}{2} - 2\mu^{(1)} \end{pmatrix} = \frac{5}{2} + 5\mu^{(0)} - 30(\mu^{(0)})^2$$

$f(\mu^{(1)})$ gets maximum for $\mu^{(1)} = \frac{1}{12}$. So
$$\mathbf{x}^{(2)} = \begin{pmatrix} 1 \\ \frac{1}{2} \end{pmatrix} + \frac{1}{12} \begin{pmatrix} 1 \\ -2 \end{pmatrix} = \begin{pmatrix} \frac{13}{12} \\ \frac{1}{3} \end{pmatrix}.$$

and
$$| \mathbf{grad} f \begin{pmatrix} \frac{13}{12} \\ \frac{1}{3} \end{pmatrix} | = 0{,}37 < 0{,}5$$

So the process breaks down at this point.

(c) The exact solution is calculated from:
$$\mathbf{grad} f(x, y) = \begin{pmatrix} 0 \\ 0 \end{pmatrix}$$

Explicitly:
$$4 - 4x + 2y = 0$$
$$2 + 2x - 12y = 0$$

The solution is
$$x = \frac{13}{11}; y = \frac{4}{11}.$$

This is to be compared with the approximate solution
$$x^{(2)} = \frac{13}{12}; y^{(2)} = \frac{1}{3} = \frac{4}{12}.$$

(d) Substitution of variables. To do this, set $y=x+3$ in $f(x, y)$ equal. This replacement leads to:
$$f(x) = -6x^2 - 24x - 48.$$

The maximum of this function is at $x_M = -2$, resubstitution yields $y_M = 1$.
The Lagrange function is:
$$\mathcal{L} = 4x - 2x^2 + 2y - 6y^2 + 2xy - \lambda(x+3-y)$$

The three conditions
$$\frac{\partial \mathcal{L}}{\partial x} = 0; \frac{\partial \mathcal{L}}{\partial y} = 0; \frac{\partial \mathcal{L}}{\partial \lambda} = 0$$

provide a system of linear equations for the three variables x, y, λ with the solution
$$x = -2, y = 1, \lambda = 14.$$

6.7 $f(x, y) = x^{-2}e^{-y} + x^2y^3 = 0$, then
$$\frac{dy}{dx} = -\frac{\frac{\partial f}{\partial x}}{\frac{\partial f}{\partial y}} = -\frac{-2x^{-3}e^{-y} + 2xy^3}{-x^{-2}e^{-y} + 3x^2y^2}.$$

6.8 The distance d of any point $P = (x, y)$ from the origin of the $x-y$ plane is: $d(x, y) = \sqrt{x^2 + y^2}$ Boundary condition

$$g(x, y) = x^2 - 3x + 3 - y$$

Lagrange function:
$$\mathcal{L} = \sqrt{x^2 + y^2} - \lambda(x^2 - 3x + 3 - y).$$

A necessary condition for the existence of an extremum is:
$$\frac{\partial \mathcal{L}}{\partial z_i} = 0, \quad z_i = x, y, \lambda$$

Appendix A

From the first equation we get

$$\frac{x}{\sqrt{x^2+y^2}} = \lambda(2x-3) \quad \text{or} \quad \lambda = \frac{x}{(2x-3)\sqrt{x^2+y^2}}.$$

Substituting into the second equation gives:

$$\frac{y}{\sqrt{x^2+y^2}} + \frac{x}{(2x-3)\sqrt{x^2+y^2}} = 0 \quad \text{or} \quad y = -\frac{x}{2x-3}.$$

Substituting into the third equation gives:

$$x^2 - 3x + 3 + \frac{x}{2x-3} = 0 \quad \text{or} \quad 2x^3 - 9x^2 + 16x - 9 = 0.$$

This equation has the only real solution $x=1$. Thus: $P=(1, 1)$ is the point of the parabola $y=x^2-3x+3$, which has the smallest distance from the origin of the coordinate system.

6.9

1. The solution of the side condition

$$g(x, y) = -\frac{3}{2}x + 3y + 6 = 0.$$

according to $y(x)$ provides

$$y = -2 + \frac{x}{2}.$$

Substitution into the function $f(x, y)$ leads to the elimination of the variable y, we obtain a function $\overline{f(x)}$ with

$$\overline{f(x)} = x^2 + 2x\left(2 + \frac{x}{2}\right) = 2x^2 - 4x.$$

Differentiation according to x results in:

$$\frac{d\overline{f(x)}}{dx} = 4x - 4 \stackrel{!}{=} 0 \quad \Longrightarrow \quad x = 1.$$

Since

$$\frac{d^2\overline{f(x)}}{dx^2} = 4 > 0$$

it is a minimum. So, the point $x=1$ is a minimum of the function $\overline{f(x)}$, since

$$y = -2 + \frac{x}{2}\bigg|_{x=1} = -\frac{3}{2}$$

we obtain that the function
$$f(x,y) = x^2 + 2xy$$
at the point $P = (1, -3/2)$ has a minimum.

2. The Lagrange function is:
$$\mathcal{L}(x, y, \lambda) = f(x,y) + \lambda \cdot g(x,y) = x^2 + 2xy + \lambda \cdot \left(-\frac{3}{2}x + 3y + 6\right).$$

The necessary condition for the occurrence of an extremum is
$$\frac{\partial \mathcal{L}}{\partial x} = \frac{\partial \mathcal{L}}{\partial y} = \frac{\partial \mathcal{L}}{\partial \lambda} = 0.$$

We obtain the following partial derivatives:
$$\frac{\partial \mathcal{L}}{\partial x} = 2x + 2y - \frac{3}{2}\lambda$$
$$\frac{\partial \mathcal{L}}{\partial y} = 2x + 3\lambda$$
$$\frac{\partial \mathcal{L}}{\partial \lambda} = -\frac{3}{2}x + 3y + 6.$$

The solution to this linear equation system is
$$x = 1, y = -3/2, \lambda = -2/3.$$

6.10 Lagrange function:
$$\mathcal{L}(r_1, r_2, \lambda) = a r_1^\alpha r_2^\beta + \lambda(b - k_1 r_1 + k_2 r_2).$$

Partial derivatives:
$$\frac{\partial \mathcal{L}}{\partial r_1} = \alpha \cdot a \cdot r_1^{\alpha-1} r_2^\beta - \lambda \cdot k_1 \stackrel{!}{=} 0$$
$$\frac{\partial \mathcal{L}}{\partial r_2} = \beta \cdot a \cdot r_1^\alpha r_2^{\beta-1} - \lambda \cdot k_2 \stackrel{!}{=} 0$$
$$\frac{\partial \mathcal{L}}{\partial \lambda} = b - k_1 r_1 + k_2 r_2 \stackrel{!}{=} 0.$$

From the first equation we obtain by solving for λ:
$$\lambda = \frac{a \cdot \alpha}{k_1} \cdot r_1^{\alpha-1} \cdot r_2^\beta,$$

from the second condition:
$$\lambda = \frac{a \cdot \beta}{k_2} \cdot r_1^\alpha \cdot r_2^{\beta-1}.$$

Appendix A

Setting these two equations equal eliminates the Lagrange multiplier λ, we get:

$$\frac{a \cdot \alpha}{k_1} \cdot r_1^{\alpha-1} \cdot r_2^{\beta} = \frac{a \cdot \beta}{k_2} \cdot r_1^{\alpha} \cdot r_2^{\beta-1}$$

or

$$r_2 = \frac{\beta}{\alpha} \cdot \frac{k_1}{k_2} \cdot r_1.$$

Inserting this relationship into the third condition results in:

$$b - k_1 \cdot r_1 - k_2 \cdot r_2 = 0$$

or

$$b - k_1 \cdot r_1 - \frac{\beta}{\alpha} k_1 \cdot r_1 = 0.$$

Solving for r_1 gives:

$$r_1 = \frac{\alpha \cdot b}{k_1(\alpha + \beta)}.$$

Inserting into the above relationship between r_1 and r_2 yields:

$$r_2 = \frac{\beta \cdot b}{k_2(\alpha + \beta)}.$$

6.11

(a) The volume of a cylinder is:

$$V(r, h) = r^2 \pi \cdot h.$$

The surface of the cylinder is:

$$O = 2 \cdot r^2 \pi + 2\pi \cdot r \cdot h.$$

The goal is to maximize the volume under the condition that the surface is $2m^2$. Therefore

$$O = 2 \cdot r^2 \pi + 2\pi \cdot r \cdot h \stackrel{!}{=} 2$$

We solve this side condition for $h(r)$ and obtain:

$$h(r) = \frac{1 - r^2 \pi}{\pi \cdot r}.$$

So:

$$V(r) = r^2 \cdot \pi \cdot h(r) = r - r^3 \cdot \pi.$$

The condition for the volume to be maximal is

$$\frac{dV(r)}{dr} = 0 \iff 1 - 3r^2\pi = 0 \implies r_m = \frac{1}{\sqrt{3\pi}}.$$

The negative solution is not meaningful and there is only the unique solution r_m.

(b) This gives

$$V(r_m) = r_m - r_m^3 \cdot \pi = \frac{2}{3} \cdot \frac{1}{\sqrt{3\pi}}.$$

The height of the cylinder with the maximum content is then

$$h(r_m) = \frac{1 - r_m^2 \cdot \pi}{\pi \cdot r_m} = 2 \cdot \frac{1}{\sqrt{\pi}}.$$

(c) In this case, the surface must be minimized under the condition that the volume is constant, or has the value 1 m³.

$$V = 1 \text{ m}^3 \implies r^2 \cdot \pi \cdot h = 1,$$

also

$$h(r) = \frac{1}{r^2\pi}.$$

If you set this value in the formula for the surface of the cylinder, you get:

$$O(r) = 2r^2 \cdot \pi + \frac{2}{r}.$$

This is minimal when

$$O'(r) = 4\pi \cdot r - \frac{2}{r^2} \stackrel{!}{=} 0.$$

The solution is

$$r_{min} = \frac{1}{(2\pi)^{1/3}}.$$

Then

$$h(r_{min}) = \frac{(2\pi)^{2/3}}{\pi},$$

and

$$O(r_{min}) = 3 \cdot (2\pi)^{1/3}$$

(d) The body with the 'more favorable' ratio is the sphere. It applies

$$V_k = \frac{4}{3}\pi \cdot r^3$$

and

$$O_k = 4\pi \cdot r^2$$

Appendix A

A sphere with a volume of 1 m³ has the radius:

$$\frac{4}{3}\pi r^3 = 1 \quad \Longrightarrow \quad r = \left(\frac{3}{4\pi}\right)^{1/3} \approx 0{,}62 \text{ m}.$$

The surface of the sphere with this radius is:

$$O_k = 4\pi \cdot \left(\frac{3}{4\pi}\right)^{2/3} \approx 4{,}83 \text{ m}^2.$$

The surface of the cylinder with the volume of one cubic meter is

$$O_z = 3 \cdot (2\pi)^{1/3} \approx 5{,}536 \text{ m}^2,$$

i.e. the sphere only requires 87.25 % of the sheet metal as the cylinder.

Chapter 7

7.1 $K_{163} = 5067{,}92 €.$.

7.2 $p = 17{,}59 \%$.

7.3 $K_0 = 9737{,}10 €$.

7.4 Linear interest rate $n = 5$ years. Compound interest $n = 4{,}81$ years.

7.5 Interest rate: 1,5 %, Interest factor: $q = 1015$, Accumulation factor for 10 years: $1{,}015^{10} = 1{,}1605$, Final capital after 10 years: $K_{10} = 9284 €$.

7.6 $K_3 = 2.091€, K_7 = 2.241€.$.

7.7 $n = 53.6$ years

7.8 (a) $A = 20.386 €$, (b) $T_{20} = 20.084 €$, (c) $K_{10} = 188.003 €$, (d) Amortization schedule:

Period	Interest	Amortization	Costs	Remaining debt
0	0	0	0	350.000,00
1	5250,00	15.136,00	20.386,00	334.864,00
2	5022,96	15.363,04	40.772,00	319.500,96
3	4792,51	15.593,49	61.158,00	303.907.47
4	4,558.61	15,827.39	81,544.00	288,080.09
5	4,321.20	16,064.80	101,930.00	272,015.29

Period	Interest	Amortization	Costs	Remaining debt
6	4,080.23	16,305.77	122,316.00	255,709.52
7	3835,64	16.550,36	142.702,00	239.159,16
8	3587,39	16.798,61	163.088,00	222.360,55
9	3335,41	17.050,59	183.474,00	205.309,96
10	3079,65	17.306,35	203,860.00	188,003.60
11	2,820.05	17,565.95	224,246.00	170,437.66
12	2,556.56	17.829,44	244.632,00	152.608,22
13	2289,12	18.096,88	265.018,00	134.511,35
14	2017,67	18.368,33	285.404,00	116.143,02
15	1742,15	18.643,85	305.790,00	97,499.16
16	1,462.49	18,923.51	326,176.00	78,575.65
17	1,178.63	19,207.37	346.562,00	59.368,28
18	890,52	19.495,48	366.948,00	39.872,81
19	598,09	19.787,91	387.334,00	20.084,90
20	301,27	20.084,73	407.720,00	0,17

(e) For $A = 14.000$ € p. a.: $n = 31,56$ years, for $A = 13.600$ € p. a.: $n = 26,7$ years, for $A = 13.000$ € p. a.: $n = 25,16$ years.

7.9 31,56 years, 39,7 years.

7.10 $i_{nom} = 0,0247$.

7.11 (a) $n = 31,74$ years, total amount $= 476.000$ €, (b) Additional repayment after 10 years $n = 26,81$ years; total amount $= 452.150$ €.

References

Alten H.-W., Djafari Naini A., Eick B., Folkerts M., Schlosser H., Schlote K.-H., Wesemüller-Kock H., Wußing H. (2014): 4000 Jahre Algebra, Geschichte Kulturen Menschen, 2. Auflage, Springer Verlag, Berlin, Heidelberg, New York.
Anderson I. (2001): A First Course in Discrete Mathematics, SUMS, Springer, London.
Apostol T. (1971): Mathematical Analysis, A Modern Approach to Advanced Calculus, Addison Wesley, Reading Massachusetts.
Arens T., Hettlich F., Karpfinger Ch., Kockelkorn U., Lichtenegger K., Stachel H. (2018): Mathematics, 4th edition, Spektrum Akademischer Verlag, Heidelberg.
Bardi J.S. (2006): The Calculus Wars, Newton, Leibniz and the greatest mathematical clash of all time, Thunder's Mouth Press, New York.
Basieux P. (2000): The Architecture of Mathematics, Thinking in Structures, Rowohlt, Hamburg.
Bosch K. (2010): Brückenkurs Mathematik, Eine Einführung mit Beispielen und Übungsaufgaben, 14. Auflage, Oldenbourg, München.
Bronstein I.N. *et al.* (2005): Taschenbuch der Mathematik, 6. Edition, Verlag Harri Deutsch.
Bröcker T. (1980): Analysis in mehreren Variablen, B.G. Teubner, Stuttgart.
Courant R. (1971a): Vorlesungen über Differential- und Integralrechnung, I. Funktionen einer Veränderlichen. Vierte Auflage, Springer Verlag, Berlin, Heidelberg, New York.
Courant R. (1971b): Vorlesungen über Differential- und Integralrechnung, II. Funktionen mehrerer Veränderlichen. Vierte Auflage, Springer Verlag, Berlin, Heidelberg, New York.
Courant R., Robbins H. (2000): Was ist Mathematik?, Fünfte, unveränderte Auflage, Springer, Berlin, Heidelberg, New York.
Dean N. (2003): Discrete Mathematics, Pearson Education, München.
Deisenroth M.P., Faisal A.A., Ong C.S. (2020): Mathematics for Machine Learning, Cambridge University Press, Cambridge (UK).
Deiser O. (2010): Introduction to Set Theory, 3. Edition, Springer, Berlin, Heidelberg.
Dieudonné J. (1985): History of Mathematics 1700–1900, Vieweg, Braunschweig.
Derbyshire J. (2006): Unknown Quantity, A Real and Imaginary History of Algebra, Penguin Books, New York.
Domschke W., Drexl A., Klein R., Scholl A. (2015): Introduction to Operations Research, 9th Edition, Springer, Berlin.
Dunham W. (1990): Journey through Genius, The Great Theorems of Mathematics. Penguin Books, New York.
Dyke P. (2018): Two and Three Dimensional Calculus with Applications in Science and Engineering. Wiley, Hoboken, NJ.

Erwe F. (1962): Differential- und Integralrechnung I, II. Bibliographisches Institut, Mannheim.
Fischer G., Springborn B. (2020): Lineare Algebra, 19., vollst. überarb. u. erg. Auflage, Springer Spektrum.
Fuchs D., Tabachnikov S. (2011): Ein Schaubild der Mathematik, Springer Verlag, Berlin.
Führer C. (2006): Kompakt-Training Wirtschaftsmathematik, Kiehl, Ludwigshafen.
Garnier R., Taylor J. (2002): Discrete Mathematics for New Technology, Bristol, Philadelphia.
Goebbels S., Ritter S. (2018): Mathematik verstehen und anwenden, 3. Auflage, Spektrum Akademischer Verlag, Heidelberg.
Graham R.L., Knuth D.E., Patashnik O. (1994), Concrete Mathematics, Second Edition, Addison-Wesley, Boston.
Gregg J. R. (1998): Ones and Zeros, Understanding Boolean Algebra, Digital Circuits and the Logic of Sets, IEEE Press, New York.
Hall R. (2002): Philosophers at War: The Quarrel Between Newton and Leibniz, Cambridge University Press.
Hilgert I., Hilgert J. (2021): Mathematics - a travel guide, 2. Edition, Springer Spektrum, Heidelberg.
Hillier F.S., Lieberman G.J. (2010): Introduction to Operations Research, Ninth Edition, MacGraw Hill, New York.
Katz V. J. (2009): A History of Mathematics, An Introduction, 3rd Edition, Addison-Wesley, Boston.
Kelly, J. (2003): Logik im Klartext, Pearson Studium, München.
Körner T.W. (2020): Where do numbers come from?, Cambridge University Press, Cambridge UK.
Koop A., Moock, H. (2018): Linear Optimization, An application-oriented introduction to Operations Research, 2. Edition, Spektrum Akademischer Verlag, Heidelberg.
Koshy T. (2001): Fibonacci and Lucas Numbers with Applications. John Wiley and Sons.
Koshy T. (2004): Discrete Mathematics with Applications, Elsevier, Amsterdam.
Kramer J., von Pippich A.-M. (2013): Von den natürlichen Zahlen zu den Quaternionen; Basiswissen Zahlbereiche und Algebra, Springer Spektrum, Wiesbaden.
Kreuzer M., Kühling S. (2006): Logik für Informatiker, Pearson Studium, München.
Lang S. (1986): A First Course in Calculus, Fifth Edition, Springer Verlag, New York.
Lang S. (1987): Linear Algebra, Third Edition, UTM, Springer Verlag, New York.
Lang S. (1988): Basic Mathematics, Springer Verlag, New York.
Maor E. (2017): To Infinity and Beyond, New Edition, Princeton University Press, Princeton, New Jersey.
Maor E. (2015): e: The Story of a Number, Princeton University Press, Princeton, New Jersey.
Maor E. (2019): The Pythagorean Theorem, A 4000-Year History, Princeton University Press, Princeton, New Jersey.
Maor E. (2020): Trigonometric Delights, 3. Edition, Princeton University Press, Princeton, New Jersey.
Marsden J., Weinstein A. (1985): Calculus I, Second Edition, Springer, New York.
Marsden J., Weinstein A. (1985): Calculus II, Second Edition, Springer, New York.
Marsden J., Weinstein A. (1985): Calculus III, Second Edition, Springer, New York.
Merzbach U.C., Boyer C.B. (2011): A History of Mathematics, Third Edition, John Wiley & Sons, Inc., Hoboken, New Jersey.
Papula L. (2018): Mathematics for Engineers and Natural Scientists, Volume 1, 15. Edition, Springer Vieweg.

References

Papula L. (2015): Mathematics for Engineers and Natural Scientists, Volume 2, 14. Edition, Springer Vieweg.

Pesic P. (2005): Abel's Proof, Springer Verlag, Berlin, Heidelberg.

Purkert W. (2014): Brückenkurs Mathematik für Wirtschaftswissenschaftler, 8. aktualisierte Auflage, Springer-Gabler.

Posamentier A. S., Lehmann I. (2007): The (Fabulous) Fibonacci Numbers, Prometheus Books, New York.

Range R.M. (2016): What is Calculus? From Simple Algebra to Deep Analysis, World Scientific Press, Singapore.

Rommelfanger H. (2010): Mathematik für Wirtschaftswissenschaftler, Band 1 und 2, 6. bzw. 5. Auflage, Spektrum Akademischer Verlag, Heidelberg.

Schäfer W., Georgi K., Otto Ch., Trippler G. (2006): Mathematik-Vorkurs, Übungs- und Arbeitsbuch für Studienanfänger, 6. Auflage, Vieweg Teubner, Stuttgart.

Spivak M. (2008): Calculus, Third Edition, Cambridge University Press, Cambridge.

Staab F. (2012): Logik und Algebra, 2. Auflage Oldenbourg.

Stillwell J. (2002): Mathematics and its History, Second Edition, Springer Verlag, New York.

Stöppler S. (1982): Mathematics for Economists, 3. Auflage, Gabler.

Tietze J. (2015): Introduction to Financial Mathematics, 12., erweiterte Auflage, Springer Spektrum.

Tietze J. (2019): Introduction to Applied Economics Mathematics, 18. Auflage, Springer Spektrum.

Toenniessen F. (2019): Das Geheimnis der transzendenten Zahlen, 2. Auflage, Spektrum Akademischer Verlag, Heidelberg.

Winter R. (2001): Grundlagen der formalen Logik, 2. überarbeitete Auflage, Verlag Harri Deutsch, Frankfurt.

Made in the USA
Monee, IL
03 May 2026

49438557R00168